MODERN ROAD

A PRACTICAL TREATISE ON THE ENGINEERING PROBLEMS
ROAD BUILDING, WITH CAREFULLY COMPILED SPECI-
FICATIONS FOR MODERN HIGHWAYS, AND
CITY STREETS AND BOULEVARDS

BY

AUSTIN T. BYRNE
CIVIL ENGINEER

AUTHOR OF
"HIGHWAY CONSTRUCTION," "MATERIALS AND WORKMANSHIP"

ILLUSTRATED

AMERICAN TECHNICAL SOCIETY
CHICAGO
1917

COPYRIGHT, 1917, BY
AMERICAN TECHNICAL SOCIETY

COPYRIGHTED IN GREAT BRITAIN
ALL RIGHTS RESERVED

INTRODUCTION

THE science of good road building is an old one as evidenced by the many highways in Europe which have withstood the wear of travel for centuries. Most of these famous roads were cut from solid rock or built of crushed stone of such a character as to be unaffected by weather conditions. Modern road building, however, has been largely influenced within the past fifteen years by the enormous increase in the amount of travel due to the automobile. This has not only been the means of developing new road surface to meet the more severe requirements of this type of vehicle but it has developed a country-wide interest in good roads, thus making it possible for the enthusiastic travelers to take long tours without meeting the formerly ever-present bugaboo of bad roads, besides making the ordinary town-to-town travel more satisfactory.

¶ It is with the idea of giving a clear conception of the engineering problems involved in road building, that is, laying out of the road by the best and easiest route, the questions of grade, contour, and drainage, and the construction of culverts and bridges, that this treatise has been written. The author has had long experience in the field of highway construction and has treated the different types of roads in a very complete and practical manner. Natural soil, gravel, broken stone, bituminous macadam and concrete roads are all carefully treated, not only as to material, but as to the best methods of laying them. The city pavements are also given due consideration, accompanied by typical specifications for the new surfaces developed for boulevards.

¶ Altogether, the article covers the entire field of road building, both city and country, and should appeal either to the highway engineer or to the untrained reader who has merely a passing interest in the subject.

VIEW OF NEW YORK STATE HIGHWAY FROM ALBANY TO SELKIRK—TYPICAL TARVIA DUSTLESS ROAD

CONTENTS

COUNTRY ROADS AND BOULEVARDS

PAGE

Resistance to movement of vehicles...................................... 1

Resistance to traction.. 1

Tractive power and gradients.. 7

Axle friction.. 11

Resistance of air.. 12

Location of roads... 12

Reconnoissance.. 13

Preliminary survey.. 15

Topography.. 15

Map... 20

Memoir.. 22

Bridge sites... 22

Final selection of route.. 22

Preliminary road construction methods................................. 35

Width and transverse contour...................................... 35

Drainage... 38

Types of drainage.. 38

Nature of soils.. 39

Location of drains.. 39

Proper fall for drains... 40

Materials used for drains.. 40

Sizes of drains.. 42

Silt basins... 42

Protection of drain ends from weather............................ 42

Drain outlets.. 43

Side ditches... 43

Treatment of springs found in cuttings........................... 44

Drainage for hillside roads...................................... 45

Inner and outer road gutters..................................... 45

Culverts... 46

Earthenware pipe culverts....................................... 49

Iron pipe culverts.. 51

Box culverts... 52

Arch culverts.. 53

Short span bridges used as culverts.............................. 53

Earthwork... 55

Balancing cuts and fills... 55

Side slopes.. 55

Shrinkage of earthwork... 57

Prosecution of earthwork.. 58

Methods of forming embankments................................. 58

CONTENTS

Preliminary road construction methods (Continued)

PAGE

Tools for construction work. 60

Natural-soil roads. 74

    Earth roads. 74

    Sand roads. 77

    Sand-clay roads. 77

    Application of oil to sand and gravel soils. 78

Roads with special coverings. 79

Foundations. 79

    Materials. 79

    Thickness. 79

    Types of foundation to be used. 81

Wearing surfaces. 82

Maintenance and improvement of roads. 109

Repair and maintenance of broken-stone roads. 109

Systems of maintenance. 110

Improvement of existing roads. 110

Traffic census. 111

CITY STREETS AND HIGHWAYS

Foundations. 121

Stone-block pavements. 123

    Materials. 124

    Cobblestone pavement. 125

    Belgian-block pavement. 125

    Granite-block pavement. 126

    Blocks. 127

    Manner of laying blocks. 127

    Foundation. 129

    Cushion coat. 129

    Laying blocks. 130

    Ramming. 130

    Fillings for joints. 130

    Stone pavement on steep grades. 132

Brick pavements. 133

    Qualifications of brick. 133

    Tests for paving brick. 136

    Brick-pavement qualifications. 137

    Foundation. 137

    Sand cushion. 137

    Manner of laying. 138

    Joint fillings. 139

    Tools used by hand in the construction of block pavements. 145

    Concrete-mixing machine. 145

    Gravel heaters. 146

CONTENTS

	PAGE
Wood-block pavements	147
Creosoting	147
Laying the blocks	149
Qualifications of wood pavements	152
Asphalt pavements	153
Sheet-asphalt pavement	153
Laying the pavement	155
Foundation	157
Qualifications of asphalt pavements	157
Failure of asphalt pavement	159
Rock asphalt pavement	161
Asphalt blocks	161
Tools employed in construction of asphalt pavements	162
Miscellaneous pavements	164
Burnt clay	164
Straw	164
Oyster-shell	164
Chert	164
Slag	164
Clinker	165
Petrolithic	165
Kleinpflaster	165
Iron	165
Trackways	165
National pavement	166
Filbered asphalt pavement	166
Miscellaneous street work	166
Curbstones and gutters	170
Curbstones	170
Combination curb and gutter	171
Street cleaning	172
Cleaning methods	172
Removal of snow	175
Street sprinkling	176
Selecting the pavement	176
Qualifications	176
Interests affected	177
Problem involved in selection	177
Economic benefit	181
Relative economies	181
Gross cost of pavements	184
Comparative rank of pavements	185
Specifications	185
Contracts	187

DISTRIBUTING TARVIA COATING ON THE WEARING COURSE OF A SURFACED MACADAM ROAD

HIGHWAY CONSTRUCTION

PART I

COUNTRY ROADS AND BOULEVARDS

RESISTANCE TO MOVEMENT OF VEHICLES

The object of a road is to provide a way for the transportation of persons and goods from one place to another with the least expenditure of power and expense. The facility with which this traffic or transportation may be conducted over any given road depends upon the resistance offered to the movement of vehicles. This resistance is composed of: (1) resistance offered by the road-way, which consists of (a) "friction" between the surface of the road and the wheel tires, (b) resistance offered to the rolling of the wheels occasioned by the want of uniformity in the road surface or lack of strength to resist the penetrating efforts of loaded wheels, (c) resistance due to gravity called "grade resistance"; (2) resistance offered by vehicles, termed "axle friction"; and (3) resistance of the air. The magnitude of each of the components has a wide range, varying with the kind and condition of the road and its surface, the form and condition of the vehicle, the load, and the speed.

Resistance to Traction. The combination of road resistances is designated by the general term "resistance to traction", the magnitude of which is measured by the number of pounds of effort per ton of the load required to overcome it; this is ascertained by a form of spring-balance variously called "dynograph", "tracto-graph", etc., one end of which is attached to the vehicles and the other end to the draft animals.

The road which offers the least resistance to traffic should combine a surface on which the friction of the wheels is reduced to the least possible amount, while possessing sufficient roughness to afford good foothold for the draft animals and good adhesion for motor vehicles; and should be so located as to give the most direct route with the least gradients.

TABLE I

Resistance to Traction on Different Road Surfaces

ROAD SURFACE	TRACTION RESISTANCE	
	Pounds per Ton	In Terms of Load
Earth road—ordinary condition	50 to 200	$\frac{1}{40}$ to $\frac{1}{10}$
Gravel	50 to 100	$\frac{1}{40}$ to $\frac{1}{20}$
Sand	100 to 200	$\frac{1}{20}$ to $\frac{1}{10}$
Macadam	30 to 100	$\frac{1}{67}$ to $\frac{1}{20}$
Plank road	30 to 50	$\frac{1}{67}$ to $\frac{1}{40}$
Steel wheelway	15 to 40	$\frac{1}{133}$ to $\frac{1}{50}$

Friction. The resistance of friction arises from the rubbing of the wheel tires against the surface of the road; its amount can be determined only by experiment for each kind of road surface. From many experiments the following deductions are drawn:

(1) The resistance to traction is directly proportional to the pressure.

(2) On solid unyielding surfaces, the resistance is independent of the width of the tire; but on compressible surfaces it decreases as the width of the tire increases. There is no material advantage gained, however, in making a tire more than 4 inches wide, for the reason that it is impossible to distribute the load evenly over the road owing to the irregularities and curvatures of its surface.

(3) On uniformly smooth surfaces, the resistance is independent of the speed.

(4) On rough irregular surfaces, which give rise to constant concussion, the resistance increases with the speed.

Table I shows the relative resistance to traction of various surfaces. These coefficients refer to the power required to keep the load in motion. It requires from two to six or eight times as much force to start a load as it does to keep it in motion at two or three miles per hour. The extra force required to start a load is due in part to the fact that during the stop the wheel may settle into the road surface; in part to the fact that the axle friction at starting is greater than after motion has begun; and in part to the fact that energy is consumed in accelerating the load.

Resistance to Rolling. Resistance to rolling is caused partly by the wheel penetrating or sinking below the surface of the road, forming a depression or rut, as shown in Fig. 1, thus compelling the wheel to be continually rolling up a short incline. The measure of this resistance is the horizontal force necessary at the axle to roll it up the incline; and is equal to the product of the load multiplied by one-third of the semi-chord of the submerged arc of the wheel.

Resistance to rolling is also caused by the wheel striking or colliding with loose or projecting stones, which suddenly checks

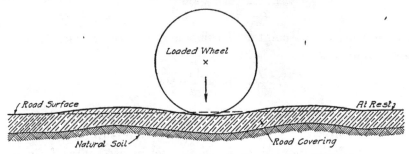

Fig. 1. Exaggerated Section of Road under Pressure of Loaded Vehicle

the motive power; the momentum thus destroyed varies with the height of the stone or obstacle and is often considerable.

In both cases the power required to overcome the resistance is affected largely by the diameter of the wheel, as the larger the wheel the less force is required to lift it over the obstruction or to roll it up the inclination due to the indentation of the surface.

Illustrative Example. The power required to draw a wheel over a stone or any obstacle, such as S in Fig. 2, may be thus calculated:

Let P represent the power sought, or that which would just balance the weight on the

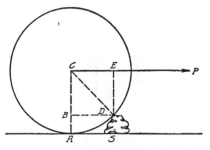

Fig. 2. Diagram for Calculating Power Required to Draw Wheel over Resisting Object

point of the stone, and the slightest increase of which would draw it over. This power acts in the direction CP with the leverage of BC or DE. The force of gravity W resists in the direction CB with the leverage BD. The equation of equilibrium will be $P \times CB = W \times BD$, whence

$$P = W\frac{BD}{CB} = W\frac{\sqrt{CD^2 - BC^2}}{CA - AB}$$

Let the radius of the wheel equal $CD = 26$ inches, and the height of the obstacle equal $AB = 4$ inches. Let the weight $W = 500$ pounds, of which 200 pounds may be the weight of the wheel and 300 pounds the load on the axle. The formula then becomes

$$P = 500\,\frac{\sqrt{676-484}}{26-4} = 500\,\frac{13.85}{22} = 314.7 \text{ lb.}$$

The pressure at the point D is compounded of the weight and the power, and equals

$$W\frac{CD}{CB} = 500 \times \frac{26}{22} = 591 \text{ lb.}$$

Therefore this pressure acts with this great effect to destroy the road in its collision with the stone; in addition there is to be considered the effect of the blow given by the wheel in descending from it. For minute accuracy the non-horizontal direction of the draft and the thickness of the axle should be taken into account. The power required is lessened by proper springs to vehicles, by enlarged wheels, and by making the line of draft ascending.

Illustrative Example. The mechanical advantage of the wheel in surmounting an obstacle may be computed from the principle of the lever. Let the wheel, Fig. 3, touch the horizontal line of traction in the point A and meet a protuberance BD. Suppose the line of draft CP to be parallel to AB. Join CD and draw the perpendiculars DE and DF. We may suppose the power to be applied at E and the weight at F, and the action is then the

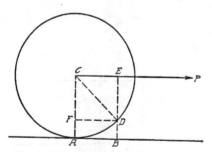

Fig. 3. Force Diagram for Wheel Drawn over Obstacle

same as the bent lever EDF turning round the fulcrum at D. Hence $P : W :: FD : DE$. But $FD : DE :: \tan FCD : 1$; and $\tan FCD = \tan 2\,DAB$; therefore $P = W \tan 2\,DAB$. Now it is obvious that the angle DAB increases as the radius of the circle diminishes; therefore, the weight W being constant, the power required to overcome an obstacle of given height is diminished when the diameter is increased. Large wheels are, therefore, the best adapted for surmounting inequalities of the road.

TABLE II

Resistance Due to Gravity on Different Inclinations

Grade 1 inch	20	30	40	50	60	70	80	90	100	200	300	400
Rise in feet per mile	264	176	132	105	88	75	66	58	52	26	17	13
Resistance in pounds per ton	100	$66\frac{2}{3}$	50	40	$33\frac{1}{3}$	$28\frac{4}{7}$	25	$22\frac{2}{9}$	20	10	$6\frac{2}{3}$	5

There are, however, circumstances which provide limits to
the height of the wheels of vehicles. If the radius AC exceeds
the height of that part of the horse to which the traces are attached,
the line of traction CP will be inclined to the horse, and part of the
power will be exerted in pressing the wheel against the ground.
The best average size of wheels is considered to be about 6 feet in
·diameter. Wheels of large diameter do less damage to a road than
small ones,' and cause less draft for the horses. With the same load,
a two-wheeled cart does far more damage than one with four wheels,
and this because of their sudden and irregular twisting motion in
the trackway.

Grade Resistance. Grade resistance is due to the action of
gravity, and is the same on good and bad roads. On level roads
its effect is immaterial, as it acts in a direction perpendicular to
the plane of the horizon and neither accelerates nor retards motion.
On inclined roads it offers considerable resistance, proportional to
the steepness of the incline. The resistance due to gravity on any
incline in pounds per ton is equal to

$$\frac{2000}{\text{rate of grade}}$$

Table II shows the resistance due to gravity on different grades.
The additional resistance caused by inclines may be investigated
in the following illustrated example.

Illustrative Example. Suppose the whole weight to be borne
on one pair of wheels, and that the tractive force is applied in a
direction parallel to the surface of the road.

Let AB, Fig. 4, represent a portion of the inclined road, C
being a vehicle just sustained in its position by a force acting in the
direction CD. It is evident that the vehicle is kept in its position
by three forces: namely, by its own weight W acting in the vertical
direction CF; by the force F applied in the direction CD parallel

to the surface of the road; and by the pressure P which the vehicle exerts against the surface of the road acting in the direction CE perpendicular to the same. To determine the relative magnitude of these three forces, draw the horizontal line AG and the vertical line BG; then, since the two lines CF and BG are parallel and are both cut by the line AB, they must make the two angles CFE and ABG equal; also the two angles CEF and AGB are equal; therefore, the remaining angles FCE and BAG are equal, and the two triangles CFE and ABG are similar. And as the three sides of the former are proportional to the three forces by which the vehicle is sustained, so also are the three sides of the latter; namely, the length of the road AB is proportional to W, or the weight of the vehicle; the vertical rise BG is proportional to F, or the force required to sustain the vehicle on the incline; and the horizontal distance AG in which the rise occurs is proportional to P, or the force with which the vehicle presses upon the surface of the road. Therefore,

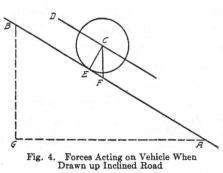

Fig. 4. Forces Acting on Vehicle When Drawn up Inclined Road

$$W : AB :: F : GB$$

and

$$W : AB :: P : AG$$

If to AG such a value be assigned that the vertical rise of the road is exactly one foot, then

$$F = \frac{W}{AB} = \frac{W}{\sqrt{AG^2 + 1}} = W \sin A$$

and

$$P = \frac{W \times AG}{AB} = \frac{W \times AG}{\sqrt{AG^2 + 1}} = W \cos A$$

in which A is the angle BAG.

To find the force requisite to sustain a vehicle upon an inclined road (the effects of friction being neglected), divide the weight of the vehicle and its load by the inclined length of the road, the vertical rise of which is one foot, and the quotient is the force required.

TABLE III

Tractive Power of Horses at Different Velocities

Miles per Hour	Tractive Force (lb.)	Miles per Hour	Tractive Force (lb.)
$\frac{3}{4}$	333.33	$2\frac{1}{4}$	111.11
1	250	$2\frac{1}{2}$	100
$1\frac{1}{4}$	200	$2\frac{3}{4}$	90.91
$1\frac{1}{2}$	166.66	3	83.33
$1\frac{3}{4}$	142.86	$3\frac{1}{2}$	71.43
2	125	4	62.50

To find the pressure of a vehicle against the surface of an inclined road, multiply the weight of the loaded vehicle by the horizontal length of the road, and divide the product by the inclined length of the same; the quotient is the pressure required. The force with which a vehicle presses upon an inclined road is always less than its actual weight; the difference is so small that, unless the inclination is very steep, it may be taken equal to the weight of the loaded vehicle.

To find the resistance to traction in passing up or down an incline, ascertain the resistance on a level road having the same surface as the incline, to which add, if the vehicle ascends, or subtract, if it descends, the force requisite to sustain it on the incline; the sum or difference, as the case may be, will express the resistance.

Tractive Power and Gradients. Although transportation by mechanically propelled vehicles will continue to increase, it is not probable that for many years it will become more important than traffic drawn by animals; and as mechanically propelled vehicles can ascend any grade feasible for animals, it is only necessary to discuss the effect of grades on horse-drawn traffic.

Tractive·Power of Horses. The necessity for easy grades is dependent upon the power of the horse to overcome the resistance to motion, which is composed of four forces, viz, friction, collision, gravity, and resistance of air. All estimates on the tractive power of horses must to a certain extent be vague, owing to the different strengths and speeds of animals of the same kind, as well as to the extent of their training to any particular kind of work.

The draft or pull which a good average horse, weighing 1,200 pounds, can exert on a level, smooth road 'at a speed of $2\frac{1}{2}$ miles per

TABLE IV
Variation of Tractive Power with Time

Hours per Day	Tractive Force (lb.)	Hours per Day	Tractive Force (lb.)
10	100	7	$146\frac{6}{7}$
9	$111\frac{1}{9}$	6	$166\frac{2}{3}$
8	125	5	200

NOTE: The tractive power of teams may be found by multiplying th
above values by the following constants:

$$\begin{aligned}
1 \text{ horse} &= 1 \\
2 \text{ horses} \quad 0.95 \times 2 &= 1.90 \\
3 \text{ horses} \quad 0.85 \times 3 &= 2.55 \\
4 \text{ horses} \quad 0.80 \times 4 &= 3.20
\end{aligned}$$

hour is 100 pounds; which is equivalent to 22,000 foot-pounds pe
minute, or 13,200,000 foot-pounds per day of 10 hours. The tractiv
power diminishes as the speed increases and, perhaps, withi
certain limits, say from $\frac{3}{4}$ mile to 4 miles per hour, nearly in invers
proportion to it. Thus the average tractive force of a horse, on
level, and actually pulling for 10 hours, may be assumed approx
imately as shown in Table III.

The work done by a horse is greatest when the velocity wit
which he moves is one-eighth of the greatest velocity with whic
he can move when unloaded; and the force thus exerted is 0.4
of the utmost force that he can exert at a dead pull. The tractiv
power of a horse may be increased in about the same proportio
as the time is diminished, so that when working from 5 to 10 hour
on a level, it will be about as shown in Table IV.

Loss of Tractive Power on Inclines. In ascending inclines
horse's power diminishes rapidly; a large portion of his strength i
expended in overcoming the resistance of gravity due to his ow
weight and that of the load. Table V shows that as the steepnes
of the grade increases, the efficiency of both the horse and the roa
surface diminishes; that the more of the horse's energy which i
expended in overcoming gravity, the less remains to overcome th
surface resistance.

Table VI shows the gross load which an average horse, weighin
1,200 pounds, can draw on different kinds of road surfaces, on
level and on grades rising 5 and 10 feet per 100 feet.

TABLE V
Effects of Grades upon the Load a Horse Can Draw on Different Pavements

Grade	Earth	Broken Stone	Stone Blocks	Asphalt
Level	1.00	1.00	1.00	1.00
1 : 100	.80	.66	.72	.41
2 : 100	.66	.50	.55	.25
3 : 100	.55	.40	.44	.18
4 : 100	.47	.33	.36	.13
5 : 100	.41	.29	.30	.10
10 : 100	.26	.16	.14	.04
15 : 100	.10	.05	.07	...
20 : 100	.04	...	.03	...

The decrease in the load which a horse can draw upon an incline is not due alone to gravity; it varies with the amount of foothold afforded by the road surface. The tangent of the angle of inclination should not be greater than the coefficient of tractional resistance. Therefore, it is evident that the smoother the road surface, the easier should be the grade; the smoother the surface the less the foothold, and consequently the less the possible load.

The loss of tractive power on inclines is greater than any investigation will show; for, besides the increase of draft caused by gravity, the power of the horse is much diminished by fatigue upon a long ascent, and even in greater ratio than man, owing to its anatomical formation and great weight. Though a horse on a level is as strong as five men, on a grade of 15 per cent, it is less strong than three; for three men carrying each 100 pounds will ascend such a grade faster and with less fatigue than a horse with 300 pounds.

A horse can exert for a short time twice the average tractive pull which he can exert continuously throughout the day's work; hence, so long as the resistance on the incline is not more than double the resistance on the level, the horse will be able to take up the full load which he is capable of drawing.

Steep grades are thus seen to be objectionable, and particularly so when a single one occurs on an otherwise comparatively level road, in which case the load carried over the less inclined portions must be reduced to what can be hauled up the steeper portion.

The bad effects of steep grades are especially felt in winter,

TABLE VI

Gross Loads for Horse on Different Pavements on Different Grades

DESCRIPTION OF SURFACE	LEVEL	GRADE (5 per cent)	GRADE (10 per cent)
Asphalt	13,216	. . . . .	. . . . .
Broken stone (best condition)	6,700	1,840	1,060
Broken stone (slightly muddy)	4,700	1,500	1,000
Broken stone (ruts and mud)	3,000	1,390	890
Broken stone (very bad condition)	1,840	1,040	740
Earth (best condition)	3,600	1,500	930
Earth (average condition)	1,400	900	660
Earth (moist but not muddy)	1,100	780	600
Stone-block pavement (dry and clean)	8,300	1,920	1,090
Stone-block pavement (muddy)	6,250	1,800	1,040
Sand (wet)	1,500	675	390
Sand (dry)	1,087	445	217

when ice covers the roads, for the slippery condition of the surface causes danger in descending, as well as increased labor in ascending; during heavy rains the water also runs down the road and gulleys it out, destroying its surface, thus causing a constant expense for repairs. The inclined portions are subject to greater wear from horses ascending, thus requiring thicker covering than the more level portions, and hence increasing the cost of construction.

It will rarely be possible, except in a flat or comparatively level country, to combine easy grades with the best and most direct route. These two requirements will often conflict. In such a case, increase the length of the road. The proportion of this increase will depend upon the friction of the covering which is adopted. But no general rule can be given to meet all cases as respects the length which may thus be added, for the comparative time occupied in making the journey forms an important element in any case which arises for settlement. Disregarding time, the horizontal length of a road may be increased to avoid a 5 per cent grade, seventy times the height.

Table VII shows, for most practical purposes, the force required to draw loaded vehicles over inclined roads. In the fifth column the length given is the length which would require the same motive power to be expended in drawing the load over it, as would be necessary to draw over a mile of the inclined road. The loads given in the sixth column are the maximum loads which average

TABLE VII

Data for Loaded Vehicles over Inclined Roads

Rate of Grade (ft. per 100 ft.)	Pressure on Plane (lb. per ton)	Tendency down Plane (lb. per ton)	Power Required to Haul 1 Ton up Plane (lb.)	Equivalent Length of Level Road (mi.)	Maximum Load Horse Can Haul (lb.)
0.00	2240	.00	45.00	1.000	6270
0.25	2240	5.60	50.60	1.121	5376
0.50	2240	11.20	56.20	1.242	4973
0.75	2240	16.80	61.80	1.373	4490
1.00	2240,	22.40	67.40	1.500	4145
1.25	*2240	28.00	73.00	1.622	3830
1.50	2240	33.60	78.60	1.746	3584
1.75	2240	39.20	84.20	1.871	3290
2.00	2240	45.00	90.00	2.000	3114
2.25	2240	50.40	95.40	2.120	2935
2.50.	2240	56.00	101.00	2.244	2725
2.75	2240	61.33	106.33	2.363	2620
3.00	2239	67.20	112.20	2.484	2486
4.00	2238	89.20	134.20	2.982	2083
5.00	2237	112.00	157.00	3.444	1800
6.00	2233	134.40	179.40	3.986	1568
7.00	2232	156.80	201.80	4.844	1367
8.00	2232	179.20	224.20	4.982	1235
9.00	2231	201.60	246.60	4.840	1125
10.00	2229	224.00	269.00	5.977	1030

* Near enough for practice; actually 2239.888

Pressure on plane = weight×nat cos of angle of plane

horses weighing 1,200 pounds can draw over such inclines, the friction of the surface being taken at $\frac{1}{50}$ of the load drawn.

Axle Friction. The resistance of the hub to turning on the axle is the same as that of a journal revolving in its bearing, and has nothing to do with the condition of the road surface. The coefficient of journal friction varies with the material of the journal and its bearing, and with the lubricant. It is nearly independent of the velocity, and seems to vary about inversely as the square root of the pressure. For light carriages when loaded, the coefficient of friction is about 0.020 of the weight on the axle; for the ordinary thimble-skein wagon when loaded, it is about 0.012. These coefficients are for good lubrication; if the lubrication is deficient, the axle friction is 2 to 6 times as much as above.

The tractive power required to overcome the above axle friction for carriages of the usual proportions is about 3 to $3\frac{3}{4}$ pounds per ton of the weight on the axle; and for truck wagons, which have medium sized wheels and axles, it is about $3\frac{1}{2}$ to $4\frac{1}{2}$ pounds per ton.

TABLE VIII

Wind Pressure for Various Vehicles

DESCRIPTION	VELOCITY OF WIND (mi. per hour)	WIND PRESSURE (lb. per sq. ft.)
Pleasant breeze	15	1.107
Brisk gale	20	1.968
	25	3.075
High wind	30	4.428
	35	6.027
Very high wind	40	7.782
	45	9.963
Storm	50	12.300

Effect of Springs on Vehicles. Experiments have shown that springs mounted in vehicles materially decrease the resistance to traction and diminish the effects caused in the vertical plane by irregularities of the surface; but they do not diminish the horizontal component which is the one that causes the greatest wear of the road, especially at speeds beyond a walking pace. The vehicles with springs were found not to cause more wear with the horses going at a trot than vehicles without springs when the horses were walking, all other conditions being similar. Vehicles with springs improperly fixed cause considerable concussion which, in turn, destroys the road covering.

Resistance of Air. The resistance arising from the force of the wind will vary with the velocity of the wind, with the velocity of the vehicle, with the area of the surface acted upon, and also with the angle of incidence of direction of the wind with the plane of the surface. Table VIII gives the force per square foot for various velocities.

LOCATION OF ROADS

The considerations governing the location of roads are dependent upon the commercial condition of the country to be traversed. In old and long-inhabited sections, the controlling element will be the character of the traffic to be accommodated. In such a section, the route is generally predetermined and, therefore, there is less liberty of choice and selection than in a new and sparsely settled district, where the object is to establish the easiest, shortest, and most economical line of intercommunication according to the physical character of the ground.

Whichever of these two cases may have to be dealt with, the same principle governs the engineer, namely, so to lay out the road as to effect the conveyance of the traffic with the least expenditure of motive power consistent with economy of construction and maintenance.

Economy of motive power is promoted by easy grades, by the avoidance of all unnecessary ascents and descents, and by a direct line; but directness must be sacrificed to secure easy grades and to avoid expensive construction.

RECONNOISSANCE

Object of Reconnoissance. The selection of the best route demands much care and consideration on the part of the engineer. To obtain the requisite data upon which to form his judgment, he must make a personal reconnoissance of the district. This requires that the proposed route be either ridden or walked over and a careful examination made of the principal physical contours and natural features of the district. The amount of care demanded and the difficulties attending the operations will altogether depend upon the character of the country. The immediate object of the reconnoissance is to select one or more trial lines, from which the final route may be ultimately determined. When there are no maps of the section traversed, or when those which can be procured are indefinite or inaccurate, the work of reconnoitering will be much increased.

Points to Consider. In making a reconnoissance there are several points which, if carefully attended to, will very considerably lessen the labor and time otherwise required. Lines which would run along the immediate bank of a large stream must of necessity intersect all the tributaries confluent on that bank, thereby demanding a corresponding number of bridges. Those, again, which are situated along the slopes of hills are more liable in rainy weather to suffer from the washing away of the earthwork and the sliding of the embankments; the others which are placed in valleys or on elevated plateaux, when the line crosses the ridges dividing the principal watercourses, will have steep ascents and descents.

In making an examination of a tract of country, the first point to attract notice is the unevenness or undulation of its surface, which appears to be entirely without system, order, or arrangement; but upon closer examination it will be perceived that one general principle of configuration obtains even in the most irregular countries. The country is intersected in various directions by main watercourses or rivers, which increase in size as they approach the point of their discharge. Towards these main rivers lesser rivers approach on both sides, running right and left through the country, and into these, again, enter still smaller streams and brooks. The streams thus divide the hills into branches or spurs having approximately the same direction as themselves, and the ground falls in every direction from the main chain of hills towards the watercourses, forming ridges more or less elevated.

The main ridge is cut down at the heads of the streams into depressions called gaps or passes; the more elevated points are called peaks. The water which has fallen upon these peaks is the origin of the streams which have hollowed out the valleys. Furthermore, the ground falls in every direction towards the natural watercourses, forming ridges more or less elevated running between them and separating from each other the districts drained by the streams.

The natural watercourses mark not only the lowest lines, but the lines of the greatest longitudinal slope in the valleys through which they flow. The direction and position of the principal streams give also the direction and approximate position of the high ground or ridges which lie between them. The positions of the tributaries to the larger stream generally indicate the points of greatest depression in the summits of the ridges and, therefore, the points at which lateral communication across the high ground separating contiguous valleys can be most readily made.

Instruments Used. The instruments employed in reconnoitering are: the *compass*, which is used to ascertain direction; the *aneroid barometer*, to fix the approximate elevation of summits, etc.; and the *hand level*, to ascertain the elevation of neighboring points. If a vehicle can be used, an odometer may be added, but distances can usually be guessed or ascertained by time estimates closely enough for preliminary purposes. The best maps obtainable and traveling companions who possess a local knowledge of the country, together with the above outfit, are all that will be necessary for the first inspection.

PRELIMINARY SURVEY

The routes selected through the reconnoissance are examined in detail by a survey called a "preliminary survey" from the results of which the exact location can be determined.

Features to Be Considered. In making the preliminary survey, the topographical features are noted to the right and left of the transit line for a convenient distance. The data required for drawing the topography are obtained by levels taken with a leveling instrument or with a transit provided with stadia wires, on lines perpendicular to the transit line of the survey. The location of buildings, fences, streams, roads, railroads, and other objects, is determined by measurements made with a tape on lines perpendicular to the survey line; or, when the distance to the object required is considerable, the location is found by angles measured from two stations on the transit line and the distance is measured by stadia. The following information is also noted: the importance, magnitude, and direction of the streams crossed; the character of the material to be excavated or available for embankments; the position of quarries; the mode of access thereto, and the kind of stone; the position of unloading points on railroads; and any other information that might affect a selection.

Topography. *Levels.* Levels should be taken along the course of each line, usually at every 100 feet, or at closer intervals, depending upon the nature of the country. In taking the levels, the heights of all existing roads, railroads, rivers, or canals should be noted. "Bench marks" should be established at least every half mile, that is, marks made on any fixed object, such as a gate post, side of a house, or, in the absence of these, a cut made on a large tree. The height and exact description of each bench mark should be recorded in the level book.

Cross Levels. Wherever considered necessary, levels at right angles to the center line should be taken. These will be found useful in showing what effect a deviation to the right or left of the surveyed line would have. Cross levels should be taken at the intersection of all roads and railroads to show to what extent, if

Fig. 5. Contour Map Used in Road Surveys

any, these levels will have to be altered to suit the levels of the proposed road.

Contours. The levels of the transit and cross lines are worked into a map that shows the irregularities of the ground with reference to its elevations and depressions. Various methods are employed for delineating these upon paper. For the purpose of the engineer the method of contours, Fig. 5, is the most serviceable, since by it the true shape of the hills and valleys can be shown.

Contours are lines drawn through the points of equal altitude; that is, every point of the ground over which a contour line passes is at a certain height above a known fixed plane called the "datum".

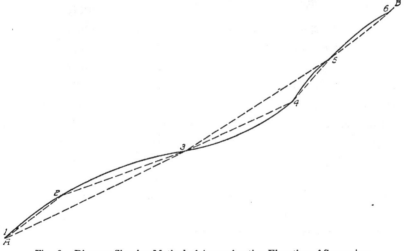

Fig. 6.   Diagram Showing Method of Approximating Elevation of Successive
Contours of Inclined Road

Mean sea level is the datum plane universally employed; when this cannot be conveniently used, an arbitrary plane is adopted which is below the lowest point in the territory under survey.

The difference of elevation between adjacent contour lines is called the "contour interval"; this may be one, five, ten, or more, feet. Whatever the difference adopted, it must be constant for all contours on the same map. Contours are designated by their height, expressed in feet, above the datum plane. The elevation of each contour is shown in figures at points close enough together to allow the eye to run from one to the other with ease. It is best to break the contour and write the numbers between the ends. If written alongside, the numbers should be placed on the higher side of the contour.

The theory of contours is given in order that no error will be made by supposing the slope of the ground from a point in one contour to a point in the next, to be a straight line. The less the contour interval, the less error will be made. If in Fig. 6 the curved line AB represents the actual surface of the ground, and points $1, 3, 5$, the elevation of successive contours, the broken line $1, 3, 5$ will represent the assumed ground surface, and its departure from the line AB is the error introduced. If now the points $2, 4, 6$ are also determined, or the contour intervals be reduced one-half, the assumed slope is $1, 2, 3, 4, 5, 6$, which differs less from the line AB than the line $1, 3, 5$, and hence introduces less error. With points determined at very short intervals, the error is practically eliminated.

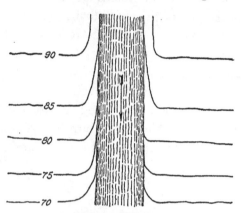

Fig. 7. Method of Showing Contours of Banks of Streams

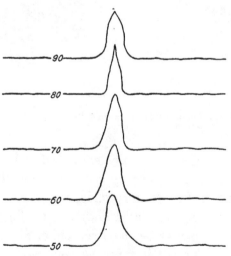

Fig. 8. Method of Showing Contour of Small Stream or Dry River Bed

A knowledge of the shape of the ground is obtained from the distance of the contours from one another. The steeper the slopes, the closer will the contours be. If in a hill the upper contours, near the summit, are closer together than those near the bottom, the intervening ground is concave; if the lower contours are closer than the higher ones, the intervening ground is convex.

Every contour must close upon itself in a loop or else must extend unbroken from one point on the margin of the map to some other point on the margin.   An exception is made in the case of large streams, the contour on each bank being carried up-stream until it cuts the water surface, when it is dropped, as shown in Fig. 7. In a small stream or dry bed, the contour crosses at the point where the elevation of the bed is that of the contour, as shown in Fig. 8.

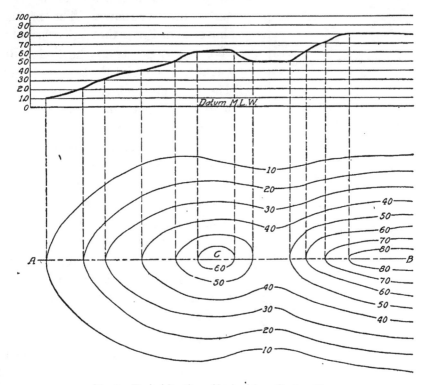

Fig. 9.   Typical Profile as Obtained from Contour Map

Profile.   A profile is a longitudinal section of the route.   The profile in any given direction is easily made from the contour map in the manner shown in Fig. 9.   Assuming that a profile is required along the line *AB*, the contours show that the ground rises from *A* to *B*, and also that a small isolated elevation occurs at *C*.   The short distance between the contours near *B* indicates that the rise is steep.   To obtain the profile, draw parallel lines at a distance apart equal to the vertical interval between the contours on any convenient exaggerated scale.   Number these lines to correspond

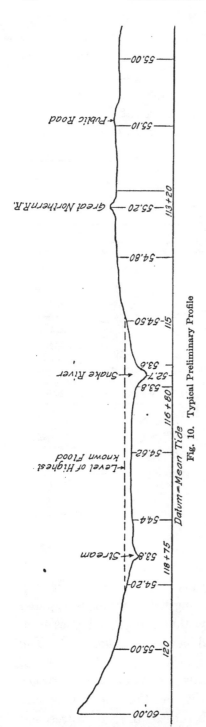

Fig. 10.   Typical Preliminary Profile

with the numbers on the contours. From each point on the line AB, where it intersects a contour, draw vertical lines to intersect the corresponding horizontal line. Connect the several points thus found, remembering the distinction between convex and concave surfaces. The profile thus obtained gives the relative heights of different points in the line AB, but it does not give the true gradient. The true gradient cannot be represented accurately unless the vertical intervals are drawn on the same scale as the horizontal scale of the map. If this is done, the elevations will generally be so minute that the profile will not give a sufficiently striking representation of the surface features. It is, therefore, necessary to exaggerate the vertical scale in a certain fixed proportion. A convenient scale is 400 feet horizontal and 40 feet vertical. A typical preliminary profile, with all the information which it is supposed to give, is shown in Fig. 10.

Map. The map, Fig. 11, should show the lengths and direction of the different portions of the line, the topography, rivers, water-courses, roads, railroads, and other matters of interest, such as town and county lines, dividing lines between property, timbered and cultivated lands, etc. Any convenient scale may be adopted; 400 feet to an inch will be found the most useful.

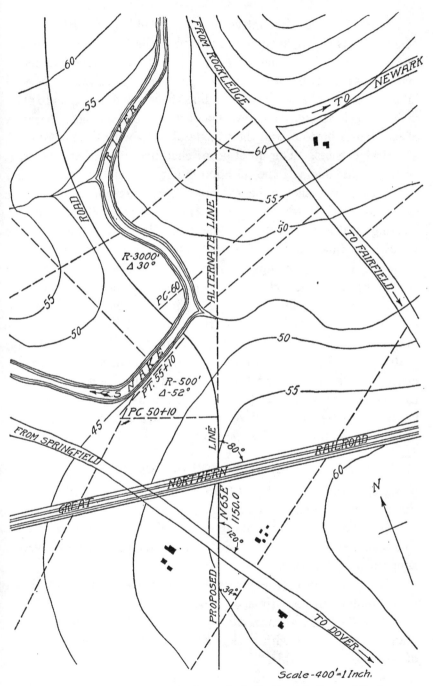

Fig. 11.   Typical Map Showing Layout in the Region of the Proposed Road

Memoir. The descriptive memoir should give with minuteness all information, such as the nature of the soil, character of the several excavations whether earth or rock, and such particular features as cannot be clearly shown upon the map or profile. Special information should be given regarding the rivers crossed, as to their width, depth at highest known flood, velocity of current, character of banks and bottom, and their angle of skew with the line of road.

Bridge Sites. The question of choosing the site of bridges is an important one. If the selection is not restricted to a particular point, the river should be examined for a considerable distance above and below what would be the most convenient point for crossing; and if a better site is found, the line of the road must be made subordinate to it. If several practicable crossings exist, they must be carefully compared in order to select the one most advantageous. The following are controlling conditions: (1) Good character of river bed, affording a firm foundation. If rock is present near the surface of the river bed, the foundation will be easy of execution, and stability and economy will be insured. (2) Stability of river banks, thus securing a permanent concentration of the waters in the same bed. (3) Axis of bridge at right angles to direction of current. (4) Bends in rivers, not being suitable localities, to be avoided if possible. A straight reach above the bridge should be secured if possible.

FINAL SELECTION OF ROUTE

Elements Entering into Choice. In making the final selection, the following principles should be observed as far as practicable:

(1) To follow that route which affords the easiest grades. The easiest grade for a given road will depend on the kind of covering adopted for its surface.

(2) To connect the places by the shortest and most direct route commensurate with easy grades.

(3) To avoid all unnecessary ascents and descents. When a road is encumbered with useless ascents, the wasteful expenditure of power is considerable.

(4) To give the center line such a position, with reference to the natural surface of the ground, that the cost of construction shall be reduced to the smallest possible amount.

(5) To cross all obstacles, where structures are necessary, as nearly as possible at right angles. The cost of skew structures increases nearly as the square of the secant of the obliquity.

(6) To cross ridges through the lowest pass which occurs.

(7) To cross either under or over railroads; for grade crossings mean danger to every user of the highway.

Treatment of Typical Cases

Connecting Two Towns in Same or Adjacent Valleys. In laying out the line of a road, there are three cases which may have to be treated, and each of these is exemplified in the contour map, Fig. 5. *First*, the two places to be connected, as the towns A and B on the plan, may both be situated in the same valley, and upon the same side of it; that is, they are not separated from each other by the main stream which drains the valley. This is the simplest case. *Second*, although both in the same valley, the two places may be on opposite sides of the valley, as at A and C, being separated by the main river. *Third*, the two places may be situated in different valleys, separated by an intervening ridge of ground more or less elevated, as at A and D. In laying out an extensive line of road, it frequently happens that all these cases have to be dealt with. The most perfect road is that of which the course is perfectly straight and the surface practically level; and, all other things being the same, the best road is that which answers nearest to this description.

Case 1. Now, in the first case, that of the two towns situated on the same side of the main valley, there are two methods which may be pursued in forming a communication between them. A road following the direct line between them, shown by the thick dotted line AB, may be made, or a line may be adopted which will gradually and equally incline from one town to another, supposing them to be at different levels; or, if they are on the same level, the line should keep at that level throughout its entire course, following all the sinuosities and curves which the irregular formation of the country may render necessary for the fulfillment of these conditions. According to the first method, a level or uniformly inclined road might be made from one to the other; this line would cross all the valleys and streams which run down to the main river, thus necessitating deep cuttings, heavy embankments, and numerous bridges; or these expensive works might be avoided by following the sinuosities of the valley. When the sides of the main valley are pierced by numerous ravines with projecting spurs and ridges intervening, instead of following the sinuosities, it will be found better to make a nearly straight line cutting through the projecting points in such a way that the material excavated should be just sufficient to fill the hollows.

Of all these, the best is the straight uniformly inclined or level road, although at the same time it is the most expensive. If the importance of the traffic passing between the places is not sufficient to warrant so great an outlay, it will become a matter of consideration whether the course of the road should be kept straight, its surface being made to undulate with the natural face of the country; or whether, a level or equally inclined line being adopted, the course of the road should be made to deviate from the direct line and follow the winding course which such a condition is supposed to necessitate.

Case 2. In the second case, that of two places situated on opposite sides of the same valley, there is, in like manner, the choice of a perfectly straight line to connect them, which would probably require a big embankment if the road were kept level; or steep inclines if it followed the surface of the country; or by winding the road, it might be carried across the valley at a higher point, where, if the level road be taken, the embankment would not be so high, or, if kept on the surface, the inclination would be reduced.

Case 3. In the third case, there is, in like manner, the alternative of carrying the road across the intervening ridge in a perfectly straight line, or of deviating it to the right and left, and crossing the ridge at a point where the elevation is less. The proper determination of the question which of these courses is the best under certain circumstances involves a consideration of the comparative advantages and disadvantages of inclines and curves. What additional increase in the length of the road would be equivalent to a given inclined plane upon it; or conversely, what inclination might be given to a road as an equivalent to a given decrease in its length? To satisfy this question, the comparative force required to draw different vehicles with given loads must be known, both upon level and variously inclined roads.

The route which will give the most general satisfaction consists in following the valleys as much as possible and rising afterward by gentle grades. This course traverses the cultivated lands, regions studded with farmhouses and factories. The value of such a line is much more considerable than that of a route by the ridges. The watercourses which flow down to the main valley are, it is true, crossed where they are the largest and require works of large dimensions, but also they are fewer in number.

Treatment of Intermediate Towns. Suppose that it is desired to construct a road between two distant towns, A and B, Fig. 12, and let us for the present neglect altogether the consideration of the physical features of the intervening country, assuming that it is equally favorable whichever line we select. Now at first sight it would appear that, under such circumstances, a perfectly straight line drawn from one town to the other would be the best that could be chosen. On more careful examination, however, of the locality, we may find that there is a third town C, situated some-what on one side of the straight line which we have

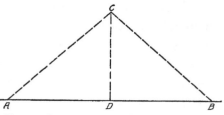

Fig. 12. Diagram Showing Method of Determining Line of Road between Successive Towns

drawn from A to B; and although our primary object is to connect only the two latter, it would nevertheless be of considerable service if all three towns were put into mutual connection with each other.

This may be effected in three different ways, any one of which might, under the circumstances, be the best. In the first place, we might, as originally suggested, form a straight road from A to B, and in a similar manner two other straight roads from A to C, and from B to C, and this would be the most perfect way of effecting the object in view, the distance between any of the two towns being reduced to the least distance possible. It would, however, be attended with considerable expense, and it would be necessary to construct a much greater length of road than according to the second plan, which would be to form, as before, a straight road from A to B, and from C to construct a road which should join the former at a point D, so as to be perpendicular to it. The traffic between

A or *B* and *C* would proceed to the point *D* and then turn off to *C*. With this arrangement, while the length of the roads would be very materially decreased, only a slight increase would be occasioned in the distance between *C* and the other two towns. The third method would be to form only the two roads *AC* and *CB*, in which case the distance between *A* and *B* would be somewhat increased, while that between *A* and *C* or *B* and *C* would be diminished, and the total length of road to be constructed also would be lessened.

As a general rule it may be taken that the last of these methods is the best and most convenient for the public; that is to say, that if the physical character of the country does not determine the course of the road, it generally will be found best not to adopt a perfectly straight line, but to vary the line so as to pass through all the principal towns near its general course.

Treatment of Mountain Roads. The location of roads in mountainous countries presents greater difficulties than in an ordinary undulating country; the same latitude in adopting undulating grades and choice of position is not permissible, for the maximum must be kept before the eye perpetually. A mountain road has to be constructed on the maximum grade or at grades closely approximating it, and but one fixed point can be obtained before commencing the survey, and that is the lowest pass in the mountain range; from this point the survey must be commenced. The reason for this is that the lower slopes of the mountain are flatter than those at their summit; they cover a larger area; and they merge into the valley in diverse undulations. Consequently, a road at the foot of a mountain may be carried at will in the desired direction by more than one route, while at the top of a mountain range any deviation from the lowest pass involves increased length of line. The engineer having less command of the ground, owing to the reduced area he has to deal with and the greater abruptness of the slopes, is liable to be frustrated in his attempt to get his line carried in the desired direction.

It is a common practice to run a mountain survey up hill, but this should be avoided. Whenever an acute-angled zigzag is met with on a mountain road near the summit, the inference to be drawn is that the line, being carried up hill, on reaching the summit was too low and the zigzag was necessary to reach the desired pass

The only remedy in such a case is a resurvey beginning at the summit and running down hill. This method requires a reversal of that usually adopted. The grade line is first staked out and its horizontal location surveyed afterwards. The most appropriate instrument for this work is a transit with a vertical circle on which the telescope may be set to the angle of the maximum grade.

Loss of Height. "Loss of height" is to be carefully avoided in a mountain road. By loss of height is meant an intermediate rise in a descending grade. If a descending grade is interrupted by the introduction of an unnecessary ascent, the length of the road will be increased, over that due to the continuous grade, by the length of the portion of the road intervening between the summit of the rise and the point in the road on a level with that rise—a length which is double that due on the gradient to the height of the rise. For example, if a road descending a mountain rises, at some intermediate point, to cross over a ridge or spur, and the height ascended amounts to 110 feet before the descent is continued, such a road would be just one mile longer than if the descent had been uninterrupted; for 110 feet is the rise due to a half-mile length at a slope of 1 : 24.

Water on Mountain Roads. Water is needed by the workmen and during the construction of the road. It is also very necessary for the traffic, especially during hot weather; and if the road exceeds 5 miles in length, provision should be made to have the water either close to or within easy reach of the road. With a little ingenuity the water from springs above the road, if such exist, can be led down to drinking fountains for men, and to troughs for animals.

In a tropical country it would be a matter for serious consideration if the best line for a mountain road 10 miles in length or upwards, but without water, should not be abandoned in favor of a worse line with a water supply available.

Halting Places. On long lines of mountain roads, halting places should be provided at frequent intervals.

Alignment of Roads. No rule can be laid down for the alignment of a road—it will depend upon both the character of its traffic and the "lay of the land". To promote economy of transportation, it should be straight; but, if straightness is obtained at the expense of easy grades that might have been obtained by deflections and increase in length, it will prove very expensive to the community that uses it.

The curving road around a hill often may be no longer than the straight one over it, for the latter is straight only with reference to the horizontal plane, while it is curved as to the vertical plane; the former is curved as to the horizontal plane, but straight as to the vertical plane. Both lines curve, and we call the one passing over the hill straight only because its vertical curvature is less apparent to our eyes. Excessive crookedness of alignment is to be avoided, for any unnecessary length causes a constant threefold waste: *first*, of the interest of the capital expended in making that unnecessary portion; *second*, of the ever recurring expense of repairing it; and *third*, of the time and labor employed in traveling over it.

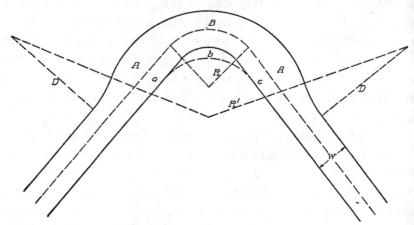

Fig. 13.   Diagram Showing Method of Laying Out Curve in Road

Location and Construction of Curves.   Curves, on a road used exclusively by horse-drawn traffic, should have a center radius of not less than 50 feet. On roads used by both horse-drawn and motor-vehicle traffic, the greatest possible radius should be employed and not less than 150 feet at the inner margin. Curves should not be placed at the foot of a steep ascent; and, when they occur on an ascent, the grade at that point should be decreased in order to compensate for the additional resistance of the curve.

Curves may be either circular or parabolic in form. The latter will be found exceedingly useful for joining tangents of unequal length and for following contours; when the curvature is least at its beginning and ending, the deviations from a straight line are less strongly marked than by a circular arc. The connection between a circular curve and its tangents should be made by a parabolic arc.

The width of the wheelway on curves should be greater than on tangents; the position in which the additional width will be of the greatest service to the traffic is at the entry arcs, as shown at A and A, Fig. 13, and not at the center B of the curve, which is the point commonly widened. The minimum radius of the outer curve to provide the increased width may be determined by the formula

$$R' = \sqrt{R + \left[\left(\frac{W+w}{\frac{1}{2}}\right)\right]^2 + l^2}$$

in which R is radius of inner curve; W is width of road on tangents; w is width of vehicles; and l is maximum length of vehicles, including teams. If the traffic requires it, a further widening may be obtained by flattening the inner curve as indicated by abc. The radius of the reversing curves should be not less than 15 feet.

The outer half of the wheelway on curves used by fast motor-vehicle traffic should be raised, as shown in Fig. 14, the amount

Fig. 14. Diagram Showing Adjustment of Profile on Curves for Rapidly Moving Motor Vehicles

of elevation being 4 inches for a curve of 150-foot radius and decreasing to 2 inches for a curve of 300-foot radius.

The approach to curves should afford an unobstructed view for at least 300 feet, and obstructions which prevent the entire length of the curve from being seen by approaching vehicles should be removed.

Zigzags. The method of surmounting a height by a series of zigzags or by a series of reaches with practicable curves at the turns, is objectionable for the following reasons:

(1) An acute-angled zigzag obliges the traffic to reverse its direction without affording it convenient room for the purpose. The consequence is that with slow traffic a single train of vehicles is brought to a stand, while if two trains of vehicles, traveling in opposite directions, meet at the zigzag, a block ensues.

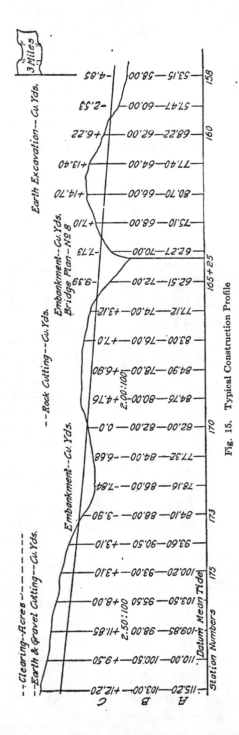

Fig. 15.　Typical Construction Profile

(2) With zigzags little progress is made toward the ultimate destination of the road; height is surmounted, but horizontal distance is increased without compensation.

(3) Zigzags are dangerous. In case of a runaway down hill, the zigzag must prove fatal.

(4) If the drainage cannot be carried clear of the road at the end of each reach, it must be carried under the road in one reach, only to appear again at the next, when a second bridge, culvert, or drain will be required, and so on at the other reaches. If the drainage can be carried clear at the termination of each reach, the lengths between the curves will be very short, entailing numerous zigzag curves, which are expensive to construct and maintain.

Details after Choosing Route

Final Location. With the route finally determined upon, it must be located. This consists in tracing the line, placing a stake at every 100 feet on the straight portions and at every 50 or 25 feet on the curves. At the tangent point of curves, and at points of compound and reverse curves, a larger and more permanent stake should be placed. Lest those stakes should be disturbed in the process of construction, their exact distance from several points outside of the ground to be occupied by the road should be carefully measured and recorded in the notebook, so that they may be replaced. The stakes above referred to show the position of the center line of the road, and form the base line from which all operations of construction are carried on. Levels are taken at each stake, and cross levels are taken at every change of longitudinal slope.

Construction Profile. The construction or working profile is made from the levels obtained on location. It should be drawn to a horizontal scale of 400 feet to the inch and a vertical scale of 20 feet to the inch. Fig. 15 represents a portion of such a profile. The figures in column A represent the elevation of the ground at every 100 feet, or where a stake has been driven, above datum. The figures in column B are the elevations of the grade above datum. The figures in column C indicate the depth of cut or height of fill; they are obtained by taking the difference between the level of the road and the level of the surface of the ground. The straight line at the top represents the grade of the road; the upper surface of the road when finished would be somewhat higher than this, while the given line represents what is termed the *sub-grade or formation level.* All the dimensions refer to the formation level, to which the surface of the ground is to be formed to receive the road covering.

At all changes in the rate of inclination of the grade line a heavier vertical line should be drawn.

Gradient. *The grade of a line is in its longitudinal slope,* and is designated by the proportion between its length and the difference of height of its two extremes. The ratio of these two qualities gives it its name; if the road ascends or falls one foot in every twenty feet of its length, it is said to have a grade of 1 : 20, or a 5 per cent grade. Grades are of two kinds: maximum and minimum. The maximum grade is the steepest which is to be permitted and which on no account is to be exceeded; the minimum grade is the least allowable for good drainage. Table IX gives different methods of designating grades.

Determination of Gradients. The maximum grade is fixed by two considerations: the one relating to the power expended in ascending, and the other to the acceleration in descending, the

TABLE IX

Methods of Designating Grades

American Method (ft. per 100 ft.)	English Method	Feet per Mile	Angle with Horizon
$\frac{1}{4}$	1 : 400	13.2	0° 8′ 36″
$\frac{1}{2}$	1 : 200	26.4	0 17 11
$\frac{3}{4}$	1 : 150	39.6	0 22 55
1	1 : 100	52.8	0 34 23
$1\frac{1}{4}$	1 : 80	66	0 42 58
$1\frac{1}{2}$	1 : $66\frac{2}{3}$	79.2	0 51 28
$1\frac{3}{4}$	1 : $57\frac{1}{4}$	92.4	1 0 51
2	1 : 50	105.6	1 8 6
$2\frac{1}{4}$	1 : $44\frac{1}{4}$	118.8	1 17 39
$2\frac{1}{2}$	1 : 40	132	1 25 57
$2\frac{3}{4}$	1 : $36\frac{1}{8}$	145.2	1 34 22
3	1 : $33\frac{1}{3}$	158.4	1 43 08
$3\frac{1}{4}$	1 : $30\frac{3}{4}$	171.6	1 51 42
$3\frac{1}{2}$	1 : $28\frac{1}{2}$	184.8	2 0 16
$3\frac{3}{4}$	1 : $26\frac{2}{3}$	198	2 8 51
4	1 : 25	211.2	1 17 26
$4\frac{1}{4}$	1 : $23\frac{1}{2}$	224.4	2 26 10
$4\frac{1}{2}$	1 : $22\frac{1}{4}$	237.6	2 34 36
$4\frac{3}{4}$	1 : 21	250.8	2 43 35
5	1 : 20	264	2 51 44
6	1 : $13\frac{2}{3}$	316.8	3 26 12
7	1 : $14\frac{2}{7}$	369.6	4 0 15
8	1 : $12\frac{1}{2}$	422.4	4 34 26
9	1 : $11\frac{1}{9}$	475.2	5 8 31
10	1 : 10	528	5 42 37

incline. There is a certain inclination, depending upon the degree of perfection given to the surface of the road, which cannot be exceeded without a direct loss of tractive power. This inclination is that, on which, in descending at a uniform speed, the traces slacken, or which causes the vehicles to press on the horses; the limiting inclination within which this effect does not take place is the *angle of repose.*

Angle of Repose. The angle of repose for any given road surface can be ascertained easily from the tractive force required upon a level with the same character of surface. Thus if the force necessary on a level to overcome the resistance of the load is $\frac{1}{40}$ of its weight, then the same fraction expresses the angle of repose for that surface.

On all inclines less steep than the angle of repose, a certain amount of tractive force is necessary in the descent as well as in the ascent, and the mean of the two drawing forces, ascending and descending, is equal to the force along the level of the road. Thus, on such inclines, as much mechanical force is gained in the descent as is lost in the ascent. From this it might be inferred that when a vehicle passes alternately each way along the road, no real loss is occasioned by the inclination of the road; which, however, is not the fact with animal power, for while the up and down slopes in the ascending journey will demand, respectively, a greater and a less number of horses than that required on a level road, no actual compensation for this fluctuation can be made in the descending journey. On inclines which are more steep than the angle of repose, the load presses on the horses during their descent, so as to impede their action, and their power is expended in checking the descent of the load; or if this effect be prevented by the use of any form of drag or brake, then the power expended on such a drag or brake corresponds to an equal quantity of mechanical power expended in the ascent, for which no equivalent is obtained in the descent.

Grade Problems. *Maximum Grade.* The maximum grade for a given road will depend: (1) upon the class of traffic that will use it, whether fast and light, slow and heavy, or mixed, consisting of both light and heavy; (2) upon the character of the pavement adopted; and (3) upon the question of cost of construction. Economy of motive power and low cost of construction are antagonistic to each other, and the engineer will have to weigh the two in the balance.

For fast and light traffic the grades should not exceed 2 per cent; for mixed traffic 3 per cent may be adopted; while for slow traffic combined with economy 5 per cent should not be exceeded. This grade is practicable but not convenient.

Minimum Grade. From the previous considerations it would appear that an absolutely level road was the one to be sought for, but this is not so; there is a minimum, or a least allowable grade, of which the road must not fall short, as well as a maximum one which it must not exceed. If the road were perfectly level in its longitudinal direction, its surface could not be kept free from water without giving it so great a rise in its middle as would expose vehicles to the danger of overturning. The minimum grade commonly used is 1 per cent.

Undulating Grades. From the fact that the power required to move a load at a given velocity on a level road is decreased on a descending grade to the same extent it is increased in ascending the same grade, it must not be inferred that the animal force expended in passing alternately each way over a rising and falling road will gain as much in descending the several inclines as it will lose in ascending them. Such is not the case. The animal force must be sufficient, either in power or number, to draw the load over the level portions and up the steepest inclines of the road, and in practice no reduction in the number of horses can be made to correspond with the decreased power required in descending the inclines.

The popular theory that a gently undulating road is less fatiguing to horses than one which is perfectly level is erroneous. The assertion that the alternations of ascent, descent, and levels, call into play different muscles, allowing some to rest while others are exerted, and thus relieving each in turn, is demonstrably false, and contradicted by the anatomical structure of the horse. Since this doctrine is a mere popular error, it should be rejected utterly, not only because it is false in itself, but still more because it encourages the building of undulating roads, and this increases the labor and cost of transportation upon them.

Level Stretches. On long ascents it is generally recommended that level or nearly level stretches be introduced at frequent intervals in order to rest the animals. These are objectionable when they cause loss of height, and animals will be more rested by halting and unharnessing for half an hour than by traveling over a level portion. The only case which justifies the introduction of levels into an ascending road is where such levels will advance the road towards its objective point; where this is the case there will be no loss of either length or height, and it will simply be exchanging a level road below for a level road above.

Establishing the Grade. When the profile of a proposed route has been made, a grade line is drawn upon it (usually in red) in such a manner as to follow its general slope, but to average its irregular elevations and depressions. If the ratio between the whole distance and the height of the line is less than the maximum grade intended to be used, this line will be satisfactory; but if it be found steeper, the cut or the length of the line will have to be increased. The latter is generally preferable.

The apex or meeting point of all curves should be rounded off by a vertical curve, as shown in Fig. 16, thus slightly changing the grade at and near the point of intersection. A vertical curve rarely need extend more than 200 feet each way from that point.

Let AB and BC be two grades in profile intersecting at station B, and let A and C be the adjacent stations. It is required to join the grades by a vertical curve extending from A to C. Imagine a chord drawn from A to C. The elevation of the middle point of the chord will be a mean of the elevations of the grade at A and C, and one-half of the difference between this and the elevation of the grade at B will be the middle ordinate of the curve. Hence we have

$$M = \frac{1}{2}\left(\frac{\text{grade } A + \text{grade } C}{2} - \text{grade } B \right)$$

in which M equals the correction in elevation for the point B. The correction for any other point is proportional to the square of its

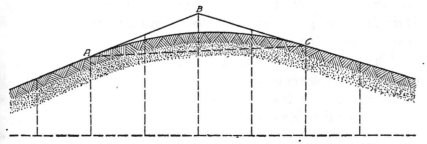

Fig. 16. Typical Road Section Showing Rounding Off of Meeting Points of Curves

distance from A to C. Thus, assuming the distance between successive ordinates, Fig. 16, as 50 feet, the correction $A+25$ is $\frac{1}{16}M$; at $A+50$ it is $\frac{1}{4}M$; at $A+75$ it is $\frac{9}{16}M$; and the same for corresponding points on the other side of B. The corrections in this case shown are subtractive, since M is negative. They are additional when M is positive, and the curve concave upward.

PRELIMINARY ROAD CONSTRUCTION METHODS
WIDTH AND TRANSVERSE CONTOUR

Width of Road. A road should be wide enough to accommodate the traffic for which it is intended, and should comprise a wheelway for vehicles and a space on each side for pedestrians.

The wheelway of country highways need be no wider than is absolutely necessary to accommodate the traffic using it; in many places a track wide enough for a single team is all that is necessary. But the breadth of the land appropriated for highway purposes should be sufficient to provide for all future increase of traffic. The wheelways of roads in rural sections should be double; that is, one portion paved (preferably the center), and the other left with the natural soil. The latter, if kept in repair, will be preferred by teamsters for at least one-half the year.

The minimum width of the paved portion, if intended to carry two lines of travel, is fixed by the width required to allow two vehicles to pass each other safely. This width is 16 feet. If intended for a single line of travel, 8 feet is sufficient, but suitable turnouts must be provided at frequent intervals. The most economical width for any roadway is some multiple of eight. Wide roads are the best; they expose a larger surface to the drying action of the sun and wind, and require less supervision than narrow ones. Their first cost is greater than that of narrow ones, and nearly in the ratio of the increased width.

The cost of maintaining a mile of road depends more upon the extent of the traffic than upon the extent of its surface, and unless-extremes be taken, the same quantity of material will be necessary for the repair of roads, either wide or narrow, which are subjected to the same amount of traffic. The cost of spreading materials over the wide road will be somewhat greater, but the cost of the materials will be the same. On narrow roads the traffic being confined to one track, will wear more severely than if spread over a wider surface.

The width of land appropriated for road purposes varies in the United States from $49\frac{1}{2}$ feet to 66 feet; in England and France from 26 to 66 feet. And the width or space macadamized is also subject to variation; in the United States the average width is 16 feet; in France it varies between 16 and 22 feet; in Belgium $8\frac{1}{4}$ feet seems to be the regular width, while in Austria, from $14\frac{1}{4}$ to $26\frac{1}{2}$ feet.

Transverse Contour. The centers of roadways in most cases should be higher than the sides, the object being to facilitate the flow of the rain water to the gutters. Where a good surface is maintained a very moderate amount of rise is sufficient for this purpose, but the rise should bear a certain proportion to the width

TABLE X

Proportionate Rise of Center to Width of Carriageway for Different Road Materials

Kind of Surface	Proportions of Rise at Center to Width of Carriageway
Earth	1:40
Gravel	1:50
Broken stone	1:60

of the carriageway. Earth roads require the most and asphalt the least. The most suitable proportions for the different paving materials is shown in Table X.

Form of Contour. All authorities agree that the form should be convex, but they differ in the amount and form of the convexity. Circular arcs, two straight lines joined by a circular arc, and ellipses, all have their advocates. For country roads a curve of suitable

Fig. 17. Typical Section of Road, Showing Contour

convexity may be obtained as follows: At ¼ of the width from center to side, make the rise ⅞ of the total rise, and at ½ of the width make the rise ⅝ of the total, Fig. 17.

Excessive height and convexity of cross section contract the width of the wheelway by concentrating the traffic at the center, that being the only part where a vehicle can run upright. The force required to haul vehicles over such cross sections is increased because an undue proportion of the load is thrown upon two wheels instead of being distributed equally over the four. The continual tread of horses' feet in one track soon forms a depression which holds water, and the surface is not so dry as with a flat section which allows the traffic to distribute itself over the whole width. Sides formed of straight lines are also objectionable. They wear hollow, retain water, and, by raising the center, defeat the object sought. The required convexity should be obtained by rounding the formation surface, and not by diminishing the thickness of the covering at the sides.

Although on hillside and mountain roads it is generally recommended that the surface should consist of a single slope inclining inwards, there is no reason for or advantage gained by this method. The form best adapted to these roads is the same as for a road under ordinary conditions.

With a roadway raised in the center and the rain water draining off to gutters on each side, the drainage will be more effectual and speedy than if the drainage of the outer half of the road has to pass over the inner half. The inner half of such a road is usually subjected to more traffic than the outer half. If formed of a straight incline, this side will be worn hollow and retain water. The inclined flat section never can be properly repaired to withstand the traffic. Consequently it never can be kept in good order, no matter how constantly it may be mended. It is always below par and when heavy rain falls it is seriously damaged.

DRAINAGE

Types of Drainage. In the construction of roads, drainage is of the first importance. The ability of earth to sustain a load depends in a large measure upon the amount of moisture retained by it. Most earths form a good firm foundation so long as they are kept dry, but when wet they lose their sustaining power, becoming soft and incoherent.

The drainage of roadways is of two kinds, viz, subsurface and surface.. The first provides for the removal of the underground water found in the body of the road; the second provides for the speedy removal of all water falling on the surface of the road. Experience has shown that a thorough removal of the underground water is of the utmost importance and is essential to the life of the road. A road covering placed on a wet undrained bottom will be destroyed by both water and frost, and will always be troublesome and expensive to maintain; perfect subsoil drainage is a necessity and will be found economical in the end even if in securing it considerable expense is required.

Subsoil Drainage

The methods employed for securing the subsoil drainage must be varied according to the character of natural soil, each kind of soil requiring different treatment.

Nature of Soils. The natural soil may be divided into the following classes: siliceous, argillaceous, and calcareous; rock, swamps, and morasses. The siliceous and calcareous soils, the sandy loams and rock, present no great difficulty in securing a dry and solid foundation. Ordinarily they are not retentive of water and therefore require no underdrains; ditches on each side of the road will generally be found sufficient. The argillaceous soils and softer marls require more care; they retain water and are difficult to compact, except at the surface; and they are very unstable under the action of water and frost.

Location of Drains. The removal of water from the subsoil is effected by drains so placed as to intercept the underground circulation of the water. Regarding the best location for the drains to accomplish this, three cases in general will present themselves:

Marginal Drains. Where the subsoil is continually wet and without a well-defined flow of water from either side. Under

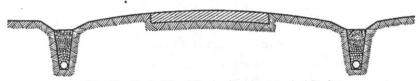

Fig. 18. Typical Road Section Showing Marginal Drains

this condition marginal drains, as shown in Fig. 18, will be found satisfactory.

Side Drains. Where there is a regular flow from one side to the other, as on a hillside road, a single drain placed on the side

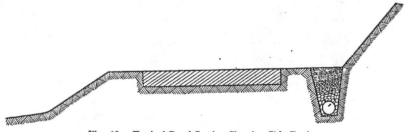

Fig. 19. Typical Road Section Showing Side Drains

from which the water comes, as in Fig. 19, will be sufficient usually.

Center and Cross Drains. Where the subsoil is so retentive of water as to require a system of drains under the roadbed, these

drains may be constructed in a variety of ways. The simples
method is to place a drain under the center of the roadway, as i
Fig. 20, connecting it at intervals by cross drains with drains place
at the sides which discharge into the natural watercourses. ·Wher

Fig. 20. Typical Road Section Showing Center Drain

the ground is level or has but a slight inclination, the cross drair
may be placed at right angles to the axis of the road. Where ther
is a steep grade it is better to lay the cross drains in the form o
an inverted V with the point in the center of the roadway an
directed uphill.

The distance apart of the cross drains depends upon the ea:
with which the subsoil yields its water. In porous soils the drair
will prove efficient at distances of from 30 to 40 feet; in retentiv
clay the spacing may range from 10 to 20 feet.

Proper Fall for Drains. The fall to be given the drair
depends upon the size of the drain and the amount of water to l
carried off. It is not advisable to employ a fall greater than 1 fo
in 100 feet. Too great a fall will produce a swift current that

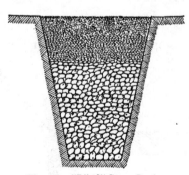

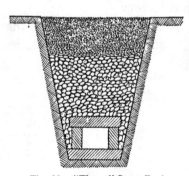

Fig. 21. "Blind" Stone Drain   Fig. 22. "Throat" Stone Drain

liable to undermine the drain as well as to choke it by foreig
matter, which a less rapid stream could not have transporte
Materials Used for Drains. The materials employed f
drains are: stone, vitrified clay pipe, porous tile, and concret

Stone Drains. The stone drain is constructed in two forms, shown in Figs. 21 and 22. The first form, called a "blind" drain, consists of a trench excavated to the required depth and filled with cobblestones or rounded pebbles. To prevent the soil from washing in and choking it, the larger stones are covered first with a layer of small gravel, and then with a layer of coarse gravel, by which means the water is filtered before passing into the porous mass beneath. Angular stones are not suitable for this type of drain. The second form of stone drain, an open channel called a "throat", is formed in the bottom of the trench with rough slabs of stone, and the trench is filled in the same manner as for a blind drain.

Vitrified Pipe Drains. Vitrified pipe drains are constructed by placing the pipe in the bottom of the trench, filling the hubs with oakum and back-filling the trench with gravel, broken stone, or a mixture of these.

Porous Tile Drains. Porous tile, Fig. 23, form very satisfactory drains. They carry off the water with great ease, rarely if ever get choked, and require only a slight inclination to keep the water moving through them. The tile have plain ends which are placed in contact in the trench and wrapped with tar paper or burlap. They are surrounded and covered with gravel or broken stone not exceeding 1 inch in

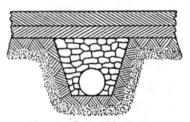

Fig. 23. Porous Tile Drain

size for a depth ranging from 6 to 12 inches, and the remaining depth of the trench is filled with large gravel or broken stone.

Concrete Tile Drains. Tile made of concrete have been in satisfactory service for several decades. They are generally made in lengths of one foot with plain ends and are laid in the same manner as the porous tile. They can be made in portable machines in the vicinity where they are to be used—an advantage that tends toward low first cost.

For the manufacture of concrete tile the best quality of hydraulic (Portland) cement, clean sand, and fine gravel or broken stone should be used in the proportion of one part cement, two parts sand, and four parts stone; the stone should contain no particles exceeding ¾ inch in size. Sufficient clean water should be used to produce a "wet mix", which should be poured into the molds before setting begins and rammed lightly. After setting is completed, the tile should be cured for about 90 days.

Sizes of Drains. The size of the drain to be adopted for a given situation depends upon the amount of water to be carried and the fall that can be given the drain. These two factors being given, there are several formulas that can be used to determine the required size. But, in the subsoil drainage of a road, the amount of water to be moved can be guessed at only; therefore, experience as to what a drain has accomplished in a given locality is a better guide than the result given by any formula. Experience shows that the least practicable size is 4 inches. The amount of water to be moved is generally assumed to vary between ¼ inch and 1 inch per acre, per 24 hours, on the area to be drained.

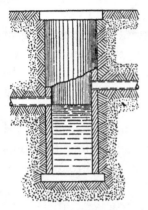

Fig. 24. Typical Construction for Silt Basin

Silt Basins. Silt basins should be constructed at all junctions and wherever else they may be considered necessary; they may be made from a single 6-inch pipe, Fig. 24, or constructed of brick masonry.

Protection of Drain Ends from Weather. As tile drains are more liable to injury from frost than those of either brick or stone,

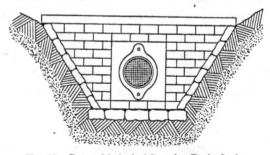

Fig. 25. Proper Method of Covering Drain Outlet

their ends at the side ditches should not be exposed directly to the weather in very cold climates, but may terminate in blind drains, or a few lengths of vitrified clay pipe reaching under the road a distance of about 3 or 4 feet from the inner slope of the ditch.

Drain Outlets. Drain outlets may be formed by building a dwarf wall of brick or stone, whichever is the cheapest or most convenient in the locality. The outlet, Fig. 25, should be covered with an iron grating to prevent vermin entering the drain pipes and building nests, thus choking the waterway.

Side Ditches. Side ditches are provided to carry away the subsoil water from the base of the road, and the rain water which falls upon its surface; to do this speedily they must have capacity and inclination proportionate to the amount of water reaching them. The width of the bed should not be less than 18 inches; the depth will vary with circumstances, but should be such that the water surface shall not reach the subgrade, but remain at least 12 inches below the crown of the road. The sides should slope at least 1½ to 1.

The longitudinal inclination of the ditch follows the configuration of the general topography, that is, the lines of natural drainage. When the latter has to be aided artificially, grades of from 1 in 500 to 1 in 800 will usually answer.

In absorbent soil less fall is sufficient, and in certain cases level ditches are permissible. The slopes of the ditches must be protected where the grade is considerable. This can be accomplished by sod revetments, riprapping, or paving.

Surface Drainage

The drainage of the roadway surface depends upon the preservation of the cross section, with regular and uninterrupted fall to the sides and without hollows or ruts in which the water can lie, and also upon the longitudinal fall of the road. If this is not sufficient the road becomes flooded during heavy rainstorms and melting snow, and is considerably damaged.

Side Ditches and Gutters. The removal of surface water from country roads may be effected by the side ditches, into which, when there are no sidewalks, the water flows directly. When there are sidewalks, gutters are formed between the roadway and footpath, and the water is conducted from these gutters into the side ditches by tile piping laid under the walks at intervals of about

50 feet. The entrance to these pipes should be protected against washing by a rough stone paving. In the case of covered ditches under the footpath, the water must be led into them by first passing through catch basins. These are small masonry vaults covered with iron gratings to prevent the ingress of stones, leaves, etc. Connection from the catch basin is made by a tile pipe about 6 inches in diameter. The mouth of this pipe is placed a few feet above the bottom of the catch basin, and the space below it acts as a depository for the silt carried by the water, and is cleaned out periodically. The catch basins may be placed from 200 to 300 feet apart. They should be made of dimensions sufficient to convey the amount of water which is liable to flow into them during heavy and continuous rains.

If on inclines the velocity of the water is greater than the nature of the soil will withstand, the gutters should be roughly paved. In all cases, the slope adjoining the footpath should be covered with sod. A velocity of 30 feet a minute will not disturb clay with sand and stone; 40 feet per minute will move coarse sand; 60 feet a minute will move gravel; 120 feet a minute should move round pebbles 1 inch in diameter, and 180 feet a minute will move angular stones $1\frac{3}{4}$ inches in diameter.

The scour in the gutters on inclines may be prevented by small weirs of stones or wood stick fascines constructed by the roadmen at a nominal cost. At junctions and crossroads the gutters and side ditches require careful arrangement so that the water from one road may not be thrown upon another; cross drains and culverts will be required at such places.

Treatment of Springs Found in Cuttings. In cuttings, springs are frequently encountered and become a source of constant danger to the stability of the slopes. In such cases the slope should be excavated at the point where the water appears, until, if possible, the source is reached. When the source has been reached, an outlet is provided by constructing a drain and connecting it with the drain at the roadside. Sometimes it may be impossible to trace the water to a single source, the whole face of the cutting being saturated for some distance. In such cases the treatment may be difficult and expensive, but a series of drains may be run up the slope to such height as will tap all the water appearing.

In cuttings, the ditch at the toe of the slope is liable to be filled with silt carried down the slope by rain; and where this might occur, covered drains should be constructed.

Drainage for Hillside Roads. On hillside or mountain roads catch-water ditches should be cut on the mountain side above the road, to cut off and convey the drainage of the ground above them to the neighboring ravines. The size of these ditches will be determined by the amount of rainfall, extent of drainage from the mountain which they intercept, and by the distances of the ravine watercourses on each side.

Inner and Outer Road Gutters. The inner road gutter should be of dimensions ample to carry off the water reaching it; when in soil, it should be roughly paved with stone. When paving is not absolutely necessary, but is desirable to arrest the scouring action of running water during heavy rains, stone weirs may be erected across the gutter at convenient intervals. The outer gutter need not be more than 12 inches wide and 9 inches deep. The gutter is formed by a depression in the surface of the road close to the parapet or revetted earthen protection mound. The drainage which falls into this gutter is led off through the parapet, or other roadside protection, at frequent intervals. The guard stones on the outside of the road are placed in and across the gutter, just below the drainage holes, so as to turn the current of the drainage into these holes or channels. On straight reaches, with parapet protection, drainage holes with guard stones should be placed every 20 feet apart. Where earthen mounds are used, and it may not be convenient to have the drainage holes or channels every 20 feet, the guardstones are to be placed in advance of the gutter to allow the drainage to pass behind them. This drainage is either to be run off at the cross drainage of the road, or to be turned off as before by a guard stone set across the gutter.

At re-entering turns, where the outer side of the road requires particular protection, guard stones should be placed every 4 feet. As all re-entering turns should be protected by parapets, the drainage holes through them may be placed as close together as desired.

Where the road is in embankment the surface water must be prevented from running down the slopes by providing ample gutters suitably connected to the natural watercourses.

Water Breaks. Water breaks to turn the surface drainage into the side ditches, should not be constructed on improved roads. They increase the grade and are an impediment to convenient and easy travel. Where it is necessary that water should cross the road, a culvert should be built.

CULVERTS

Functions of Culverts. Culverts are necessary for carrying the cross streams under a road, and also for conveying the surface water collected in the side ditches from the upper side to that side on which the natural watercourses lie.

Especial care is required to provide an ample way for the water to be passed. If the culvert is too small, it is liable to cause a washout, entailing interruption of traffic and cost of repairs, and possibly may cause accidents that will require payment of large sums for damages. On the other hand, if the culvert is made unnecessarily large, the cost of construction is needlessly increased.

Factors Considered in Design of Culverts. The area of water way required depends upon a number of important factors, which will be discussed briefly.

Rate of Rainfall. It is the maximum rate of rainfall during the severest storms which is required in this connection. This varies greatly in different sections of the country.

The maximum rainfall as shown by statistics is about one inch per hour (except during heavy storms); equal to 3,630 cubic feet per acre. Owing to various causes, not more than 50 to 75 per cent of this amount will reach the culvert within the same hour

Inches of rainfall$\times 3,630$ $=$ cubic feet per acre

Inches of rainfall$\times 2,323,200 =$ cubic feet per square mile

Kind and Condition of Soil. The amount of water to be drained off will depend upon the permeability of the surface of the ground which will vary greatly with the kind of soil, the degree of saturation, the condition of the cultivation, the amount of vegetation, etc.

Character and Inclination of Surface. The rapidity with which the water will reach the watercourse depends upon whether the surface is rough or smooth, steep or flat, barren or covered with vegetation, etc.

Condition and Inclination of Stream Bed. The rapidity with which the water will reach the culvert depends upon whether there is a well-defined and unobstructed channel or whether the water finds its way in a broad, thin sheet. If the watercourse is unobstructed and has a considerable inclination, the water may arrive at the culvert nearly as rapidly as it falls; but if the channel is obstructed, the water may be much longer in passing the culvert than in falling.

Shape of Area to be Drained and Position of Stream Branches. The area of waterway depends upon the amount of the area to be drained; but in many cases the shape of this area and the position of the branches of the stream are of more importance than the amount of the territory. For example, if the area is long and narrow, the water from the lower portion may pass through the culvert before that from the upper end arrives; or, on the other hand, if the upper end of the area is steeper than the lower, the water from the former may arrive simultaneously with that from the latter. Again, if the lower part of the area is supplied better with branches than the upper portion, the water from the former will be carried past the culvert before the arrival of that from the latter; or, on the other hand, if the upper part is supplied better with branch watercourses than is the lower, the water from the whole area may arrive at the culvert at nearly the same time. In large areas the shape of the area and the position of the watercourses are very important considerations.

Mouth of Culvert and Inclination of Bed. The efficiency of a culvert may be increased very materially by arranging the upper end so that the water may enter into it without being retarded. The discharging capacity of a culvert can be increased greatly by increasing the inclination of its bed, provided the channel below will allow the water to flow away freely after having passed the culvert.

Provision for Discharge of Water Under Head. The discharging capacity of a culvert can be increased greatly by allowing the water to dam up above it. A culvert will discharge twice as much under a head of four feet as under a head of one foot. This can be done safely only with a well-constructed culvert.

The determination of the values of the different factors entering into the problem is almost wholly a matter of judgment. An estimate for any one of the above factors is liable to be in error from 100 to 200 per cent, or even more, and of course any result deduced from such data must be very uncertain. Fortunately, mathematical exactness is not required by the problem nor warranted by the data. The question is not one of 10 or 20 per cent of increase; for if a 2-foot pipe is insufficient, a 3-foot pipe probably will be the next size, an increase of 225 per cent; and if a 6-foot arch culvert is too small, an 8-foot will be used, an increase of 180 per cent. The real question is whether a 2-foot pipe or an 8-foot arch culvert is needed.

Valuable data on the proper size of any particular culvert may be obtained as follows: (1) by observing the existing openings on the same stream; (2) by measuring, preferably at time of high water, a cross section of the stream at some narrow place; and (3) by determining the height of high water as indicated by drift and débris, and from the evidence of the inhabitants of the neighborhood.

On mountain roads, or roads subjected to heavy rainfall, culverts of ample dimensions should be provided wherever required, and it will be more economical to construct them of masonry. In localities where boulders and débris are likely to be washed down during wet weather, it will be a good precaution to construct catch pools at the entrance of all culverts and cross drains for the reception of such matter. In hard soil or rock these catch pools will be simple well-like excavations, with their bottoms two or three feet below the entrance sill or floor of the culvert or drain. Where the soil is soft they should be lined with stone laid dry; if very soft, with masonry. The size of the catch pools will depend upon the width of the drainage works. They should be wide enough to prevent the drains from being injured by falling rocks and stones of a not inordinate size.

The use of catch pools obviates the necessity of building culverts and drains at an angle to the axis of the road. Oblique structures are objectionable, as being longer than if set at right angles and by reason of the acute- and obtuse-angled terminations to their piers, abutments, and coverings.

Types of Culverts

General Classification. Three types of culverts are employed, namely: pipe, box, and arch. The pipe culvert is employed for small streams, in sizes from 12 to 24 inches. Box culverts are employed in sizes from 24 inches up to 8 feet. Arch culverts are used for spans 8 feet and over. In the construction of culverts a variety of materials are used. Pipe culverts are constructed of earthenware or vitrified clay, cast iron, corrugated steel, brick, or concrete; box and arch culverts are built of stone, brick, or concrete. Short span concrete bridges are also often employed as culverts. The type of culvert and the material to be used are determined in some cases by the cost; in others by the load to be supported, as where the depth of fill over the culvert is considerable, or where a large area of waterway is required.

Earthenware Pipe Culverts. *Construction.* In laying the pipe the bottom of the trench should be rounded out to fit the lower half of the body of the pipe, with proper depressions for the sockets. If the ground is soft or sandy, the earth should be rammed carefully, but solidly, in and around the lower part of the pipe. The top surface of the pipe, as a rule, never should be less than 18 inches below the surface of the roadway, but there are many cases where pipes have stood for several years, under heavy loads, with only 8 to 12 inches of earth over them. No danger from frost need be apprehended, provided the culverts are so constructed that the water is carried away from the level end. Ordinary soft drain tiles are not affected in the least by the expansion of frost in the earth around them.

The freezing of water in the pipe, particularly if more than half full, is liable to burst it; consequently the pipe should have a sufficient fall to drain itself, and the outside should be so low that there is no danger of backwaters reaching the pipe. If properly drained, there is no danger from frost.

Jointing. In many cases, perhaps in most, the joints are not calked. If this is not done, there is danger of the water being forced out of the joints and of washing away the soil from around the pipe. Even if the danger is not very imminent, the joints of the larger pipes, at least, should be calked with hydraulic cement, since the cost is very small compared with the insurance against damage thereby secured. Sometimes the joints are calked with clay. Every culvert should be built so it can discharge water under a head without damage to itself.

Use of Bulkheads. Although often omitted, the end sections should be protected with a masonry, Fig. 26, or timber bulkhead.

The foundation of the bulkhead should be deep enough not to be disturbed by frost. In constructing the end wall, it is well to increase the fall near the outlet to allow for a possible settlement

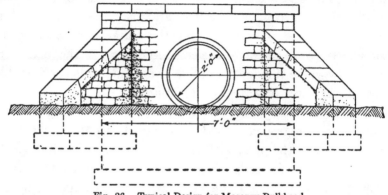

Fig. 26.   Typical Design for Masonry Bulkhead

of the interior sections.   When stone and brick abutments are too expensive, a fair substitute can be made by setting posts in the ground and spiking plank to them.   When planks are used, it is best to set them with considerable inclination towards the road-bed to prevent their being crowded outward by the pressure of the embankment.   The upper end of the culvert should be so protected

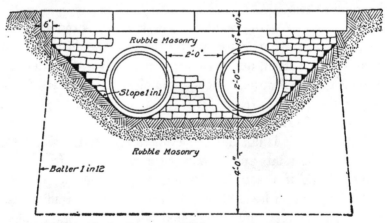

Fig. 27.   Section Showing Typical Layout for Double Pipe Culvert

that the water will not readily find its way along the outside of the pipes, in case the mouth of the culvert should become submerged.

When the capacity of one pipe is not sufficient, two or more may be laid side by side as shown in Fig. 27.   Although the two

small pipes do not have as much discharging capacity as a single
large one of equal cross section, yet there is an advantage in laying
two small ones side by side, since the water need not rise so high
to utilize the full capacity of the two pipes as would be necessary
to discharge itself through a single one of large size.

Iron Pipe Culverts. During recent years iron pipe, Fig. 28,
has been used for culverts on many prominent railroads, and may
be used on roads in sections where other materials are unavailable.

In constructing a culvert with cast-iron pipe the points requiring
particular attention are: (1) tamping the soil tightly around the
pipe to prevent the water from forming a channel along the outside;

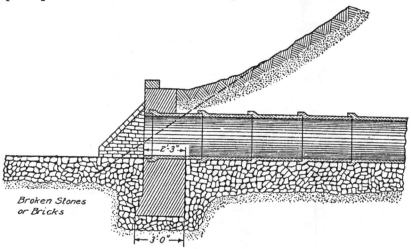

Broken Stones
or Bricks

Fig. 28. Section Showing Construction of Iron Pipe Culvert

and (2) protecting the ends by suitable head walls and, when necessary, laying riprap at the lower end. The amount of masonry
required for the end walls depends upon the relative width of the
embankment and the number of sections of pipe used.   For example,
if the embankment is, say, 40 feet wide at the base, the culvert
may consist of three 12-foot lengths of pipe and a light end wall
near the toe of the bank; but if the embankment is, say, 32 feet
wide, the culvert may consist of two 12-foot lengths of pipe and a
comparatively heavy end wall well back from the toe of the bank.
The smaller sizes of pipe usually come in 12-foot lengths, but sometimes a few 6-foot lengths are included for use in adjusting the
length of the culvert to the width of the bank. The larger sizes
are generally 6 feet long.

Box Culverts. Box culverts, Fig. 29, consist of two side walls with a flat deck. When stone is used, they are generally built of dry rubble masonry. The walls should be well founded at about 42 inches below the bed of the stream. The thickness of the walls varies according to the height. The wings are formed by extending the walls out straight and stepping them down. The deck may be made of stone slabs or reinforced concrete; with the latter it is possible to use wider spans than with stone slabs. Where the force of the stream is sufficient to scour the bed, it will be necessary to

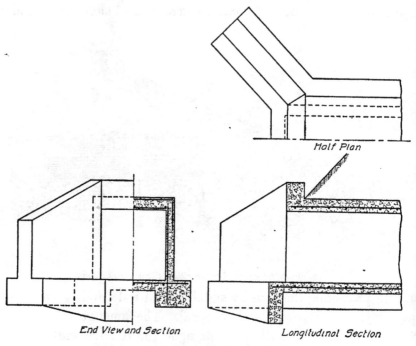

Half Plan

End View and Section   *Longitudinal Section*

Fig. 29. Typical Design of Concrete Box Culvert

pave it with stone or concrete. When reinforced concrete is used instead of stone, the side walls are made from 4 to 8 inches thick, depending upon the height. Where it is not necessary to pave the stream bed, the walls are carried down about 2 feet below the bed, and founded upon a footing 9 to 12 inches thick and sufficiently wide to secure ample area of the soil to support the load. Where scouring of the bed is liable to occur, a concrete bottom is constructed throughout the entire width and length of the culvert, and the side walls are founded on it; if necessary, a cut-off wall is constructed across each end to a depth of about 2 feet below the bottom.

Arch Culverts. The arch form of culvert is more costly than the other forms, but it is often preferred on account of its appearance, Fig. 30. When masonry and plain concrete are used, very heavy abutments are required in order that no movement can take place under a live load, to cause bending moments in the arch. In designing reinforced-concrete arches, bending is provided for by consider-

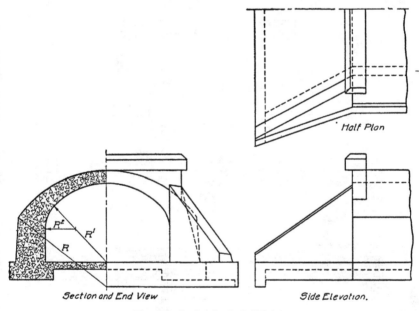

Half Plan

R^2 R'

R

Section and End View

Side Elevation.

Fig. 30. Design for Arch Culvert

ing the arch as a curved beam, with a consequent reduction in the weight of the abutments.

Short Span Bridges Used as Culverts. Three types of reinforced-concrete bridges are employed for short spans: (a) the flat slab; (b) the T-beam; (c) the steel I-beam incased in concrete, Fig. 31. The length of span over which reinforced slabs may be built with safety depends upon the load to be carried; under normal conditions the maximum span is 12 feet. The thickness of the slab for a span 2 feet should be not less than 6 inches and should increase with increase of span. The slabs are reinforced with steel bars, expanded metal, or other forms of reinforcing metal; the cross-sectional area of the reinforcing steel required is about 1 per cent of that of the slab.*

The T-beam type is practicable for spans from 12 to 30 feet. The I-beam type may be used for all spans up to 30 feet. In this type, the I-beam is designed to transmit the load to the abutments, while the reinforced-concrete floor transmits the load to the I-beams. This type of construction is noted for its safety and ability to withstand severe and unfavorable conditions, such as the settlement of the abutments, which may cause rupture of the concrete. The

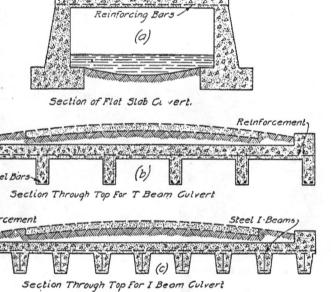

Fig. 31. Sections of Typical Short Span Concrete Bridges Used as Culverts

I-beam may or may not be incased in the concrete; the object sought in so doing is to protect it from rust. This may be accomplished also by painting, but as this needs to be repeated frequently and as there is a possibility that it will not be done, it is better to incase the beams in the concrete during construction, and so insure their permanent protection. This type also admits of arch construction between the beams for the floor system, thus decreasing the depth required for the floor; this feature may be of value in locations where the area of the waterway or the "head room" is a controlling factor.

*The theory of design of concrete bridges and culverts is discussed in Masonry and Reinforced Concrete, Part III.

EARTHWORK

The term "earthwork" is applied to all the operations performed in the making of excavations and embankments. In its widest sense it comprehends work in rock as well as in the looser materials of the earth's crust.

Balancing Cuts and Fills. In the construction of new roads, the formation of the roadbed consists in bringing the surface of the ground to the adopted grade. This grade should be established so as to reduce the earthwork to the least possible amount, both to render the cost of construction low, and to avoid unnecessarily marring the appearance of the country in the vicinity of the road. The most desirable position of the grade line is usually that which makes the amounts of cutting and filling equal to each other, for any surplus embankment over cutting must be made up by borrowing, and surplus cutting must be wasted; both of these operations involving additional cost for labor and land.

Side Slopes. *Inclination.* The proper inclination for the side slopes of cutting and embankments depends upon the nature of the soil, the action of the atmosphere, and the action of internal moisture upon it. For economy the inclination should be as steep as the nature of the soil will permit.

The usual slopes in cuttings are:

Solid rock	$\frac{1}{4}$	to 1
Earth and gravel	$1\frac{1}{2}$	to 1
Clay	3 or 6 to 1	
Fine sand	2 or 3 to 1	

The slopes of embankment are usually made $1\frac{1}{2}$ to 1.

Form of Slopes. The natural, strongest, and ultimate form of earth slopes is a concave curve, in which the flattest portion is at the bottom. This form is very rarely given to the slopes in constructing them; in fact, the reverse is often the case, the slopes being made convex, thus saving excavation by the contractor and inviting slips.

In cuttings exceeding 10 feet in depth the forming of concave slopes will aid materially in preventing slips, and in any case they will reduce the amount of material which eventually will have to be removed when cleaning up. Straight or convex slopes will continue to slip until the natural form is attained.

A revetment or retaining wall at the base of a slope will save excavation.

In excavations of considerable depth, and particularly in soils liable to slips, the slope may be formed in terraces, the horizontal offsets or benches being made a few feet in width with a ditch on

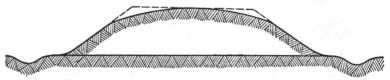

Fig. 32.　Section Showing Correct Slopes of Embankments

the inner side to receive the surface water from the portion of the side slope above them.　These benches catch and retain earth

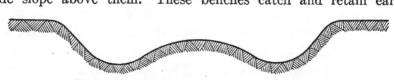

Fig. 33.　Section Showing Correct Slopes of Excavations

that may fall from the slopes above them.　The correct forms for the slopes of embankment and excavation are shown in Figs. 32 and 33.

Covering of Slopes. It is not usual to employ any artificial means to protect the surface of the side slopes from the action of the weather; but it is a precaution which in the end will save much labor and expense in keeping the roadways in good order.　The simplest means which can be used for this purpose consist in covering the slopes with good sods, or else with a layer of vegetable mold about four inches thick, carefully laid and sown with grass seed.　These means are amply sufficient to protect the side slopes from injury when they are not exposed to any other cause of deterioration than the wash of the rain and the action of frost on the ordinary moisture retained by the soil.

A covering of brushwood or a thatch of straw may also be used with good effect; but from their perishable nature they will require frequent renewal and repairs.

Where stone is abundant a small wall of stone laid dry may be constructed at the foot of the slopes to prevent any wash from them being carried into the ditches.

Shrinkage of Earthwork. All materials when excavated increase in bulk, but after being deposited in banks subside or shrink (rock excepted) until they occupy less space than in the pit from which excavated.

Rock, on the other hand, increases in volume by being broken up, and does not settle again into less than its original bulk. The increase may be taken at 50 per cent.

The shrinkage in the different materials is about as follows:

Gravel	8 per cent
Gravel and sand	9 per cent
Clay and clay earths	10 per cent
Loam and light sandy earths	12 per cent
Loose vegetable soil	15 per cent
Puddled clay	25 per cent

Thus an excavation of loam measuring 1000 cubic yards will form only about 880 cubic yards of embankment, or an embankment of 1000 cubic yards will require about 1120 cubic yards, measured in excavation, to make it. A rock excavation measuring 1000 yards will make from 1500 to 1700 cubic yards of embankment, depending upon the size of the fragments.

The lineal settlement of earth embankments will be about in the ratio given above; therefore either the contractor should be instructed, in setting his poles to guide him as to the height of grade on an earth embankment, to add the required percentage to the fill marked on the stakes, or the percentage may be included in the fill marked on the stakes. In rock embankments this is not necessary.

Classification of Earthwork. Excavation is usually classified as earth, hardpan, loose rock, or solid rock. For each of these classes a specific price is usually agreed upon, and an extra allowance is sometimes made when the haul, or distance to which the excavated material is moved, exceeds a given amount.

The characteristics which determine the classes to which a given material belongs are usually described with clearness in the specifications, as:

Earth, to include loam, clay, sand, and loose gravel.

Hardpan, to include cemented gravel, slate, cobbles, and boulders containing less than 1 cubic foot, and all other material of an earthy nature, however compact it may be.

Loose rock, to include shale, decomposed rock, boulders, and detached masses of rock containing not less than 3 cubic feet, and all other material of a rock nature which may be loosened with pick, although blasting may be resorted to in order to expedite the work.

Solid rock, to include all rock found in place in ledges and masses, or boulders measuring more than 3 cubic feet, and which can only be removed by blasting.

Prosecution of Earthwork. No general rule can be laid down for the exact method of carrying on an excavation and disposing of the excavated material. The operation in each case can be determined only by the requirements of the contract, character of the material, magnitude of the work, length of haul, etc.

Methods of Forming Embankments. *General Case.* When embankments are to be formed less than 2 feet in height, all stumps, weeds, etc., should be removed from the space to be occupied by the embankment. For embankments exceeding 2 feet in height stumps need only be close cut. Weeds and brush, however, ought to be removed and if the surface is covered with grass sod, it is advisable to plow a furrow at the toe of the slope. Where a cut passes into a fill all the vegetable matter should be removed from the surface before placing the fill. The site of the bank should be examined carefully and all deposits of soft, compressible matter removed. When a bank is to be made over a swamp or marsh the site should be drained thoroughly, and if possible the fill should be started on hard bottom.

Perfect stability is the object aimed at, and all precautions necessary to this end should be taken. Embankments should be built in successive layers: banks 2 feet and under in layers from inches to 1 foot; heavier banks in layers 2 and 3 feet thick. The horses and vehicles conveying the materials should be required to pass over the bank for the purpose of consolidating it, and care should be taken to have the layers dip towards the center. Embankments which have been first built up in the center, and afterwards widened by dumping the earth over the sides, should never be allowed.

Embankments on Hillsides. When the axis of the road is laid out on the side slope of a hill, and the road is formed partly

by excavating and partly by embanking, the usual and most simple
method is to extend out the embankment gradually along the whole
line of the excavation. This method is insecure; the excavated
material if simply deposited on the natural slope is liable to slip,
and no pains should be spared to give it a secure hold, particularly
at the toe of the slope. The natural surface of the slope should be
cut into steps, as shown in Fig. 34. The dotted line AB represents
the natural surface of the ground, CEB the excavation, and ADC
the embankment, resting on steps which have been cut between A
and C. The best position for these steps is perpendicular to the
axis of greatest pressure. If AD is inclined at the angle of repose
of the material, the steps near A should be inclined in the opposite direction to AD, and at an angle of nearly 90 degrees thereto,

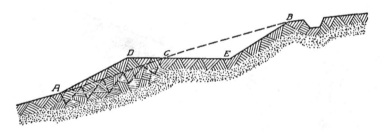

Fig. 34. Section of Embankment Showing Reinforcing by Means of Steps

while the steps near C may be level. If stone is abundant in the
locality, the toe of the slope may be further secured by a dry wall
of stone.

On hillsides of great inclination the above method of construction will not be sufficiently secure; retaining walls of stone must
be substituted for the side slopes of both the excavations and embankments. These walls may be made of stone laid dry, when stone
can be procured in blocks of sufficient size to render this kind of
construction of sufficient stability to resist the pressure of the
earth. When the stones laid dry do not offer this security, they
must be laid in mortar. The wall which forms the slope of the
excavation should be carried up as high as the natural surface
of the ground. Unless the material is such that the slope may be
safely formed into steps or benches, as shown in Fig. 34, the wall
that sustains the embankment should be built up to the surface

of the roadway, and a parapet wall or fence raised upon it, to protect pedestrians against accident, Fig. 35.

For the formula for calculating the dimensions of retaining walls see Instruction Paper on Masonry and Reinforced Concrete, Part III.

Treatment of Roadways on Rock Slopes. On rock slopes, when the inclination of the natural surface is not greater than 1 on the vertical to 2 on the base, the road may be constructed partly in excavation and partly in embankment in the usual manner, or by cutting the face of the slope into horizontal steps with vertical faces, and building up the embankment in the form of a solid stone wall in horizontal courses, laid either dry or in mortar. Care is required in proportioning the steps, as all attempts to lessen the

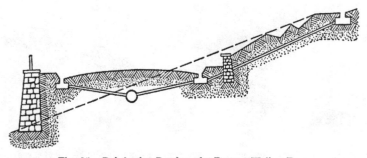

Fig. 35. Reinforcing Roadway by Parapet Wall or Fence

quantity of excavation, by increasing the number and diminishing the width of the steps, require additional precautions against settlement in the built-up portion of the roadway.

When the rock slope has a greater inclination than 1:2 the whole of the roadway should be in excavation.

In some localities roads have been constructed along the face of nearly perpendicular cliffs, on timber frameworks consisting of horizontal beams firmly fixed at one end by being let into holes drilled in the rock, the other end being supported by an inclined strut resting against the rock in a shoulder cut to receive it. There are also examples of similar platforms suspended instead of being supported.

Tools for Construction Work

Picks. Picks are made in various styles, according to the class of material in which they are to be used. Fig. 36 shows the form usually employed in street work. Fig. 37 shows the form generally used for clay or gravel excavation.

Fig. 36. Grading Pick
Courtesy of Acme Road Machinery Company, Frankfort, New York

The eye of the pick is formed generally of wrought iron, while the points are of steel. The weight of picks ranges from 4 to 9 pounds.

Fig. 37. Clay Pick
Courtesy of Acme Road Machinery Company, Frankfort, New York

Grubbing Tools. In handling brush, stumps, etc., such tools as the bush hooks, Fig. 38, the bush mattock, Fig. 39, and the axe mattock, Fig. 40, are commonly used. These are cutting as well as grading tools.

Shovels. Shovels, Fig. 41, are made in two forms, square and round pointed, usually of pressed steel.

Plows. Plows are employed extensively in grad-

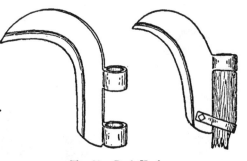

Fig. 38. Bush Hooks

Fig. 39. Bush Mattock

Fig. 40. Axe Mattock

ing, special forms being manufactured for the purpose. They are known as "grading plows", "road plows", "township plows", etc.

They vary in form according to the kind of work they are intended for, viz, loosening earth, gravel, hardpan, and some of the softer rocks.

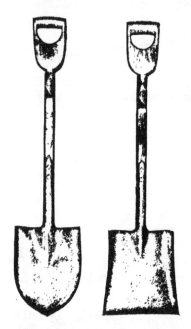

Fig. 41. Round Pointed and Square
Shovels
*Courtesy of Acme Road Machinery
Company, Frankfort, New York*

These plows are of great strength; selected white oak, rock elm, wrought steel, and iron generally being used in their construction. The cost of operating plows ranges from 2 to 5 cents per cubic yard, depending upon the compactness of the soil. The quantity of material loosened will vary from 2 to 5 cubic yards per hour.

Grading Plow. Fig. 42 shows the form usually adopted for loosening earth. This plow does not turn the soil, but cuts a furrow about 10 inches wide and of a depth adjustable up to 11 inches.

In light soil the plows are operated by 2 or 4 horses; in heavy soil as many as 8 are employed. Grading plows vary in weight from 100 to 325 pounds.

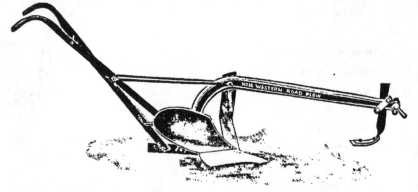

Fig. 42. Typical Road Plow
Courtesy of Western Wheeled Scraper Company, Aurora, Illinois

Hardpan Plow. Fig. 43 illustrates a plow specially designed for tearing up macadam, gravel, or similar material. The point is a straight bar of cast steel drawn down to a point, and can be repaired easily.

Scrapers. Scrapers are used generally to move the material loosened by plowing; they are made of either iron or steel, and in a

Fig. 43. Typical Hardpan or Rooter Plow
Courtesy Western Wheeled Scraper Company, Aurora, Illinois

variety of forms, and are known by various names, as "drag", "buck", "pole", and "wheeled". The drag scrapers are employed usually on short hauls, the wheeled ones on long hauls.

Drag Scrapers. Drag scrapers, Fig. 44, are made in three sizes. The smallest, for one horse, has a capacity of 3 cubic feet; the others, for two horses, have a capacity of 5 to 7½ cubic feet. The smallest weighs about 90 pounds, and the

Fig. 44. Drag Scraper
Courtesy Western Wheeled Scraper Company, Aurora, Illinois

larger ones from 94 to 102 pounds.

Buck Scrapers. Buck scrapers, Fig. 45, are made in two sizes— two-horse, carrying 7½ cubic feet; four-horse, 12 cubic feet.

Pole Scrapers. Pole scrapers are designed for use in making and leveling earth roads and for cutting and cleaning ditches; they are well adapted also for moving earth short distances at a minimum cost.

Fig. 45. Buck Scraper
Courtesy Western Wheeled Scraper Company, Aurora, Illinois

Wheeled Scrapers. Wheeled scrapers, Fig. 46, consist of a metal box, usually steel, mounted on wheels, and furnished with levers for raising, lowering, and dumping. They are operated in the same

Fig. 46. Typical Wheeled Scraper
Courtesy Western Wheeled Scraper Company, Aurora, Illinois

manner as drag scrapers, except that all the movements are made by means of the levers, and without stopping the team. By their use the excessive resistance to traction of the drag scraper is avoided.

Fig. 47. Contractor's Barrow with Pressed-Steel Tray
Courtesy of Acme Road Machinery Company, Frankfort, New York

Various sizes are made, ranging in capacity from 10 to .17 cubic feet. In weight they range from 350 to 700 pounds.

Wheelbarrows. Wheelbarrows sometimes are constructed of wood and are employed most commonly for earthwork. Their capacities range from 2 to $2\frac{1}{2}$ cubic feet. Weight is about 50 pounds.

The barrow, Fig. 47, has pressed-steel tray, oak frame, and steel wheels, and will be found more durable in the maintenance department than the all-wood barrow. Capacity is from $3\frac{1}{2}$ to 5 cubic feet, dependent on size of tray.

The barrow, Fig. 48, is constructed with tubular-iron frames and steel tray, and is adaptable to the heaviest work, such as mov-

Fig. 48. All Steel and Iron Concrete Barrow
Courtesy of Acme Road Machinery Company, Frankfort, New York

ing heavy broken stone, etc., or it may be employed with advantage in the cleaning department. Capacity from 3 to 4 cubic feet. Weight from 70 to 82 pounds.

The maximum distance to which earth can be moved economically in barrows is about 200 feet. The wheeling should be per-

Fig. 49. Typical Dump Carts for Hauling Earth, Etc.
Courtesy of Western Wheeled Scraper Company, Aurora, Illinois

formed upon planks, whose steepest inclination should not exceed 1 in 12. The force required to move a barrow on a plank is about $\frac{1}{25}$ part of the weight; on hard dry earth, about $\frac{1}{14}$ part of the weight.

The time occupied in loading a barrow will vary with the character of the material and the proportion of wheelers to shovelers. Approximately, a shoveler takes about as long to fill a barrow with earth as a wheeler takes to wheel a full barrow a distance of about 100 or 120 feet on a horizontal plank and return with the empty barrow.

Fig. 50.   Rear View of Typical Dump Wagon Showing Bottom Open
Courtesy of Western Wheeled Scraper Company, Aurora, Illinois

Carts. The cart usually employed for hauling earth, etc., is shown in Fig. 49. The average capacity is 22 cubic feet, and the average weight is 800 pounds. These carts are furnished usually

Fig. 51.   Twenty-Yard Dump Car
Courtesy of Western Wheeled Scraper Company, Aurora, Illinois

with broad tires, and the body is balanced so that the load is evenly divided about the axle.

The time required to load a cart varies with the material. One shoveler will require about as follows: clay, 7 minutes; loam, 6 minutes; sand, 5 minutes.

Dump Wagons. The use of dump wagons, Fig. 50, for moving excavated earth, etc., and for transporting materials such as sand, gravel, etc., materially shortens the time required for unloading the ordinary form of contractor's wagon; having no reach or pole connecting the rear axle with the center bearing of the front axle, they may be cramped short and the load deposited just where required. They are operated by the driver, and the capacity ranges from 35 to 45 cubic feet.

Fig. 52. Typical Grader
Courtesy of Acme Road Machinery Company, Frankfort, New York

Dump Cars. Dump cars, Fig. 51, are made to dump in several different ways, viz, single or double side, single or double end, and rotary or universal dumpers.

Dump cars may be operated singly or in trains, as the magnitude of the work may demand. They may be moved by horses or small locomotives. They are made in various sizes, depending upon the gage of the track on which they are run. A common gage is 20 inches, but it varies from that up to the standard railroad gage of $56\frac{1}{2}$ inches.

Mechanical Graders. Mechanical graders are used extensively in the making and maintaining of earth roads. They excavate and move earth more expeditiously and economically than can be

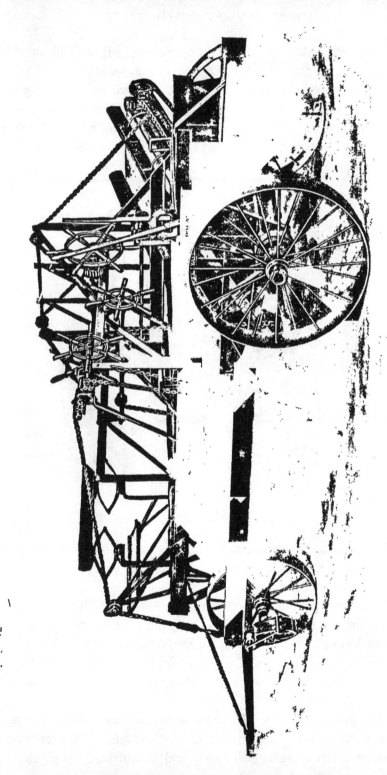

Fig. 53. Standard Elevating Grader
Courtesy of Western Wheeled Scraper Company, Aurora, Illinois

---

done by hand; they are called by various names, such as "road machines", "graders", "road hones", etc.

Simple Graders. Briefly described, graders consist of a large blade, Fig. 52, made entirely of steel, iron, or wood shod with steel, which is so arranged by a mechanism attached to the frame from which it is suspended that it can be adjusted and fixed in any direction by the operator. In their action they combine the work of excavating and transporting the earth. They have been employed chiefly in the forming and maintenance of earth roads, but also may be used advantageously in preparing the subgrade surface of roads for the reception of broken stone or other improved covering.

Elevating Graders. Some graders combine the function of elevating the material, of excavating it from side ditches, and of loading it automatically into carts or wagons. Briefly described, the machine, Fig. 53, consists of a plow which loosens and raises the earth, depositing it upon a transverse carrying belt, which conveys it from excavation to embankment. Carrier frames of two or three different lengths are provided with the machine, the distance of the end of the elevator from the plow varying from 15 to 30 feet. The carrier belt is of heavy 3-ply rubber 3 feet wide.

The plow and carrier are supported by a strong trussed framework resting on heavy steel axles and broad wheels. The large rear wheels are ratcheted upon the axle, and connected with strong gearing which propels the carrying belt at right angles to the direction in which the machine is moving.

The wheels and trusses are low and broad, occupying a space 8 feet wide and 14 feet long, exclusive of the side carrier. This enables it to work on hillsides where any wheeled implements can be used. Notwithstanding its large size it is so flexible that it may be turned around on a 16-foot embankment. Pilot wheels and levers enable the operator to raise or lower the plow or carrier at pleasure.

For motive power, 12 horses—8 driven in front, 4 abreast, and 4 in the rear on a push cart—are usually employed.

When the teams are started, the operator lowers the plow and throws the belting into gear, and as the plow raises and turns the earth to the side the belt receives and delivers it at the distance for which the carrier is adjusted, forming either excavation or embankment, as the case may be.

When it becomes necessary to deliver the excavated earth beyond the capacity of the machine, the earth is loaded upon wagons,

Fig. 54. Two Views of Elevating Graders Loading Earth into Dump Wagons
Courtesy of Western Wheeled Scraper Company, Aurora, Illinois

then conveyed to any distance. By adjusting the height of the carrier, the wagons are driven under it, Fig. 54, and loaded with 1¼ to 1½ yards of earth in from 20 to 30 seconds. When one wagon turns out with its load, another drives under the carrier, and the machine thus loads 600 to 800 wagons per day. It is claimed that with six teams and three men it is capable of excavating and placing in embankment from 1,000 to 1,500 cubic yards of earth in 10 hours, or of loading from 600 to 800 wagons in the same time, and that the cost of this handling is from 1½ to 2½ cents per cubic yard.

Points to be Considered in Selecting a Road Machine. In the selection of a road machine the following points should be carefully considered: thoroughness and simplicity of its mechanical construction; material and workmanship used in its construction; safety to the operator; ease of operation; lightness of draft; and adaptability to general road work, ditching, etc.

Care of Road Machines. The road machine when not in use should be stored in a dry house and thoroughly cleaned, its blade brushed clean from all accumulations of mud, wiped thoroughly dry, and well covered with grease or crude oil. The axles, journals, and wearing parts should be kept well oiled when in use, and an extra blade should be kept on hand to avoid stopping the machine while the dulled one is being sharpened.

Surface Graders. The surface grader is used for removing earth previously loosened by a plow. It is operated by one horse.

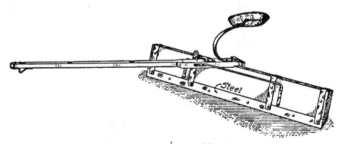

Fig. 55. Simple Road Leveler

The load may be retained and carried a considerable distance, or it may be spread gradually as the operator desires. It is also employed to level off and trim the surface following the scrapers.

The blade is of steel, ¼ inch thick, 15 inches wide, and 30 inches long. The beam and other parts are of oak and iron. Weight about 60 pounds.

Road Leveler. The road leveler, Fig. 55, is used for trimming and smoothing the surface of earth roads. It is largely employed in the spring when the frost leaves the ground.

The blade is of steel, $\frac{1}{4}$-inch thick by 4 inches by 72 inches, and is provided with a seat for the driver. It is operated by a team of horses. Weight about 150 pounds.

Ditching Tools. The tools employed for digging the ditches and shaping the bottom to fit the drain tiles are shown in Fig. 56. They are convenient to use, and expedite the work by avoiding unnecessary excavation.

The tools are used as follows: Nos. 3, 4, and 5 are used for digging the ditches; Nos. 6 and 7 for cleaning and rounding the

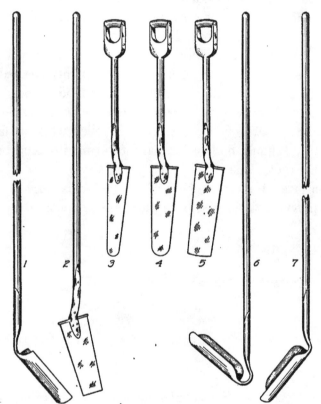

Fig. 56.   Typical Tools Used for Digging Ditches

bottom of the ditch for round tile; No. 2 is used for shoveling out loose earth and leveling the bottom of the ditch; No. 1 is used for the same purpose when the ditch is intended for "sole" tile.

Sprinkling Wagons. A convenient form of sprinkling wagon for suburban streets and country roads is shown in Fig. 57. The tank is of 12 gage steel and its capacity is 380 to 600 gallons.

Road Rollers. *Horse-Drawn Rollers.* There are a number of types of horse-drawn rollers on the market, consisting essentially

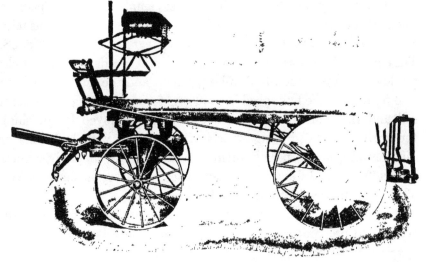

Fig. 57. Steel Tank Sprinkling Wagon
Courtesy of Acme Road Machinery Company, Frankfort, New York

of a hollow cast-iron cylinder 4 to 5 feet long, 5 to 6 feet in diameter, and weighing from 3 to 6 tons. Some forms are

Fig. 58. Ten-Ton Steam-Driven Road Roller
Courtesy of Charles Longenecker & Company, New York City

provided with boxes in which stone or iron may be placed to increase the weight, and some have closed ends and may be filled with water or sand. The use today of small gasoline road rollers makes this type less prevalent than formerly.

Power-Propelled Rollers. The rollers employed for compacting the natural soil and all forms of broken-stone pavements usually are of the three-wheel type, operated by steam or gasoline, Fig. 58. They generally are arranged to move at two speeds, low and high; the low speed is from 2 to 3 miles per hour and the high speed from 4 to 5 miles. The low speed is employed for compacting the natural soil and the foundation; the high speed is employed for finishing the surface. The driving wheels are furnished with lock pins or differential gears to permit them to accommodate themselves automatically to the difference in speeds when operating on sharp curves. They vary in weight from 10 to 20 tons.

Scarifiers. The implement used for breaking up a broken-stone road preparatory to applying a new surface is called a "scari-

Fig. 59. Scarifier, for Quick and Economical Repair of Macadam Roads
Courtesy of Charles Longenecker & Company, New York City

fier", Fig. 59. It usually consists of a cast-iron block, weighing about 3 tons, mounted on 2 or 4 wheels; the block is fitted with a series of spikes or picks, arranged either in one line, or in two lines forming a V; means are provided for adjusting the depth to which the picks penetrate, the maximum depth being about 6 inches. The scarifier is operated by being attached to the rear of a steam roller or traction engine which hauls it over the road.

NATURAL=SOIL ROADS

Earth Roads. The term "earth road" is applied to roads where the surface consists of the native soil; this class of road is the most common and cheapest in first cost. At certain seasons of the year earth roads, when properly cared for, are second to none, but during the spring and wet seasons they are very deficient in the important requisite of hardness, and are almost impassable.

For the construction of new earth roads, all the principles previously discussed relating to alignment, grades, drainage, width, etc., should be followed carefully. The crown or transverse contour should be greater than in stone roads; 12 inches at the center in 25 feet will be sufficient.

Drainage is especially important, because the material of the road is more susceptible to the action of water, and more easily destroyed by it than are the materials used in the construction of the better class of roads. When water is allowed to stand upon the road, the earth is softened, the wagon wheels penetrate it, and the horses' feet mix and knead it until it becomes impassable mud. The action of frost is also apt to be disastrous upon the more permeable surface of the earth road, having the effect of swelling and heaving the roadway and throwing its surface out of shape. It may in fact be said that the whole problem of the improvement and maintenance of ordinary country roads is one of drainage.

In the preparation of the wheelway all stumps, brush, vegetable matter, rocks, and boulders should be removed from the surface and the resulting holes filled in with clean earth. The roadbed, having been brought to the required grade and crown, should be thoroughly rolled; all inequalities appearing during the rolling should be filled up and re-rolled.

Care of Earth Roads. If the surface of the roadway is properly formed and kept smooth, the water will be shed into the side ditches and do comparatively little harm; but if it remains upon the surface, it will be absorbed and convert the road into mud. All ruts and depressions should be filled up as soon as they appear. Repairs should be attended to particularly in the spring. At that season the judicious use of a road machine and rollers will make a smooth road. In summer when the surface gets rough it can be improved by running a harrow over it; if the surface is a little muddy this treatment will hasten the drying.

During the fall the surface should be repaired, with special reference to putting it in shape to withstand the ravages of winter.

Saucer-like depressions and ruts should be filled up with clean earth similar to that of the roadbed and tamped into place.

The side ditches should be examined in the fall to see that they are free from dead weeds and grass, and late in winter they should be examined again to see that they are not clogged. The mouths of culverts should be cleaned of rubbish and the outlet of tile drains opened. Attention to the side ditches will prevent overflow and washing of the roadway, and also will prevent the formation of ponds at the roadside and the consequent saturation of the roadbed.

Holes and ruts should not be filled with stone, bricks, gravel, or other material harder than the earth of the roadway as the hard material will not wear uniformly with the rest of the road, but produce bumps and ridges, and usually result in making two holes, each larger than the original one. It is bad practice to cut a gutter

Fig. 60. Steel Road Drag
Courtesy of Western Wheeled Scraper Company, Aurora, Illinois

from a hole to drain it to the side of the road. Filling is the proper course, whether the hole is dry or contains mud.

The maintaining of smooth surfaces on all classes of earth roads will be assisted and cheapened greatly by the frequent use of a roller (either steam or horse) and any one of the various forms of road grading and scraping machines. In repairing an earth road the plow should not be used. It breaks up the surface which has been compacted by time and travel.

In the maintenance of earth roads the road drag, Fig. 60, or some similar device, is indispensable. The drag should be light and should be hauled along the road at an angle of about 45 degrees, so that only a small amount of earth is pushed to the center of the road. The driver should ride on the drag and not drive faster than a walk. Dragging should begin on the side of the road, or wheel track, and return on the opposite side. Unless the road is in good condition, it should be dragged after every heavy rain.

In the maintenance of clay roads neither sods nor turf should be used to fill holes or ruts; for, though at first deceptively tough, they soon decay and form the softest mud. Neither should the ruts be filled with field stones; they will not wear uniformly with the rest of the road, but will produce hard ridges.

Trees and close hedges should not be allowed within 200 feet of a clay road. It requires all the sun and wind possible to keep its surface in a dry and hard condition.

Sand Roads. The aim in the improvement of sand roads is to have the wheelway as narrow and well defined as possible, so as to have all vehicles run in the same track. An abundant growth of vegetation should be encouraged on each side of the wheelway, for by this means the shearing of the sand is, in a great measure, avoided. Ditching beyond a slight depth to carry away the rain water is not desirable, for it tends to hasten the drying of the sands, which is to be avoided. Where possible the roads should be overhung with trees, the leaves and twigs of which, catching on the wheelway, will serve still further to diminish the effect of the wheels in moving the sands about. If clay can be obtained, a coating 6 inches thick will be found a most efficient and economical improvement. A coating of 4 inches of loose straw will, after a few days' travel, grind into the sand and become as hard and firm as a dry clay road.

Sand=Clay Roads. A sand-clay road is formed by mixing clay and sand in such proportions that the clay will just fill the voids in the sand, and produce a mixture that is neither sticky nor friable, but coheres in a comparatively dry plastic mass when compacted with pressure. If an insufficient amount of clay is used, the mixture will not bind; if an excess of clay is used, the road will be sticky and muddy after a fall of rain.

The grains of sand furnish the hard material to resist the abrasion of the traffic; the clay provides the cementing or binding medium to hold the sand together. All clays are not equally satisfactory as binders, owing to the diversity of their origin. A common test for clay suitable for road purposes is to apply a wet finger to a piece of clay; if the clay adheres to the finger, it may be assumed reasonably that it will adhere to the sand.

The natural sand soils and the naturel clay soils are improved by the application of the sand-clay mixture, the method of applying it being varied according to which kind of soil is to be treated.

Applying Sand-Clay Mixture to Clay Soil. In the treatment of a clay soil, the soil is plowed to a depth of 6 to 8 inches; then pulverized by harrowing, and, if necessary, by rolling with a light roller and again harrowing. After the clay is thoroughly pulverized, the sand is spread over the surface in a layer from 6 to 8 inches thick, and the sand and clay are thoroughly mixed by continued harrowing. After the dry mixing has been accomplished satisfactorily, the surface is moistened slightly by sprinkling with water, then compacted by rolling, after which a road machine or grader is used to give the required crown; and then the roller is again applied until the surface becomes smooth and hard.

Applying Sand-Clay Mixture to Sand Soil. In the treatment of a sand soil, the clay is spread over the surface in a layer, ranging from 4 to 8 inches thick; then mixed with the sand by harrowing. After that it is sprinkled heavily with water and again worked with the harrow; then it is shaped and rolled in the same manner as stated above for a clay soil.

The sand-clay roads require considerable attention, after completion, to eliminate weak or defective spots by applying sand or clay, as may be required.

Application of Oil to Sand and Gravel Soils. Sand and gravel soils are improved by the application of crude petroleum or asphaltic oils. The oil abates dust; forms a non-absorbent surface which turns off rain water and decreases the amount of mud; and furnishes a dark-colored road surface which is more pleasing to the eye than the ordinary light color.

The roadbed is prepared to receive the oil by grading, shaping, and rolling. The oil is applied to the prepared surface by sprinkling from tank wagons; the oil coat is covered with a thin layer of sand, after which the roller is applied again. If during the rolling the surface becomes sticky, or dry and dusty, dry sand or more oil is added as required.

ROADS WITH SPECIAL COVERINGS

Elements of a Road Covering. The wheelways of roads and streets are prepared for traffic by placing upon the natural soil a covering of some suitable material which will furnish a comparatively smooth surface on which the resistance to traction will be reduced to the least possible amount, and over which all classes of vehicles may pass with safety and expedition at all seasons of the year. The covering usually consists of two parts: a foundation, and a wearing surface.

The functions of the foundation are as follows: (1) to protect the soil from disturbance and the injurious effects of surface water; (2) to transmit to and distribute over a sufficiently large area of the soil the weight of the loads imposed upon the wearing coat; (3) to support unyieldingly the wearing surface and the loads coming upon it.

The efficiency of the wearing surface depends entirely upon the quality of the foundation. If the foundation be weak, the wearing surface will be disrupted speedily, no matter how well constructed.

FOUNDATIONS

Materials. The foundation, when once constructed, should not require to be disturbed nor reconstructed. The materials employed in its construction may be the cheapest available, such as local rock, gravel, sand, furnace slag, etc., the important point in the design being to provide sufficient thickness, so that when consolidated it will maintain its form under the heaviest traffic liable to come upon it. If the foundation and the covering yield under the load, an upheaval is caused that disrupts the bond and hastens the destruction of the road.

Thickness. The thickness of the foundation depends upon the supporting power of the natural soil and the weight of the loads coming upon the wearing surface. The supporting or bearing power of the soil can be ascertained by direct test, and the weight of the loads by a survey of the traffic plus a provision for future increase.

Recent tests indicate that non-porous soils from which the subsoil water is removed by drainage will support in their worst condition a load of about 4 pounds per square inch; and that if the thickness of the foundation be adjusted to the traffic on this basis it will be safe at all seasons of the year.

Methods of Calculating Thickness of Covering. There are two theories as to the manner in which pressure of a loaded wheel is transmitted from the surface of the covering to the natural soil: (1) that the pressure on the soil varies inversely as the cube of the thickness of the foundation and the wearing surface; (2) that the pressure is transmitted downwards in the form of a truncated cone, the lines of which diverge at an angle varying from 30 to 50 degrees from the vertical, according to the solidity of the covering. If the surface of the road is uneven or obstructed by loose stones, the lines of pressure are more concentrated when the wheels pass over such obstacles.

The latter theory is the one most frequently applied. The calculation is performed as follows: Let P be distributed pressure on the soil, per square inch; A, length of arc of wheel tire in contact with surface in square inches; W, width of tire in inches; L, load carried by wheel in pounds; F, depth of wearing surface and foundation in inches; C, area of contact equal to $A \times W$; and B, area of base at surface of natural soil. The area of the base is

$$B = (2F + A) \ (2F + W)$$

The distributed pressure is

$$P = \frac{L}{(2F + A) \ (2F + W)} = \frac{L}{B}$$

Assuming that the load is 1,000 pounds per inch of tire width; the tire, 3 inches; length of contact 3 inches; total thickness of the wearing coat and the foundation 12 inches; the pressure on the soil is

$$P = \frac{1000 \times 3}{(2 \times 12 + 3) \ (2 \times 12 + 3)} = \frac{3000}{27 \times 27} = \frac{3000}{729} = 4.10 \text{ lb. per sq. in.}$$

According to this theory the thickness of the covering varies from 4 to 16 inches, the smallest thickness being placed upon gravel or sand and the greatest upon clay.

Preparation of Foundation. The preparation of the foundation involves two distinct operations: (1) preparation of the natural soil; and (2) placing an artificial foundation upon the prepared natural soil.

The essentials necessary to the preparation of the natural soil are: (1) the entire removal of perishable vegetable and yielding matter; (2) the drainage of the soil where necessary; (3) the improving of the bearing power of the soil where required; and (4) compacting the soil.

All soils are improved by rolling, and weak spots, which otherwise would pass unnoticed, are discovered. However, care must be taken that the weight of the roller employed is not too great for the bearing power of the soil; if it exceed this the surface of the soil will be formed into a series of undulations that will cause the wearing coat to fail; the same condition may be produced by excessive rolling with a comparatively light roller. Each soil requires different treatment.

Soils of a siliceous and calcareous nature may be improved by drainage and the addition of a layer of clay 2 to 6 inches thick, mixed with the soil and compacted by rolling. The argillaceous and allied soils, owing to their retentive nature, are very unstable under the action of water and frost, and in their natural condition afford a defective foundation. They are improved by thorough drainage and the admixture of sand well rolled, together with the placing upon the surface of the compacted soil a layer 2 to 6 inches thick of sand, slag, cinders, or other material of a similar nature, and then compacting it by sprinkling with water and rolling.

Types of Foundation to Be Used. The essential requisite in the construction of the artificial foundation is that it be a dense mass, and the type of foundation to be employed varies with the character of the wearing surface. For the various types of broken-stone surfaces, the foundation may be composed of blocks of stone (ledge rock or fieldstones), roughly shaped to a rectangular form, ranging in width and depth from 6 to 8 inches and in length from 6 to 16 inches. They are set by hand on the soil bed with the length at right angles to the axis of the roadway, so arranged that they break joints. The edges that project above the subgrade level are broken off with hand hammers, and the spaces between them are filled with chips of stone well packed and wedged in. The blocks are brought then to a firm bearing by rolling with a steam roller, after which the wearing surface is laid. The foundation also may be composed of broken stone, gravel, or furnace slag so graded that the voids will be reduced to the smallest possible amount. The voids may be filled with stone dust; a mixture of sand and clay; a mortar and grout composed of hydraulic cement and sand; bituminous cement; or hydraulic-cement concrete, mixed and placed upon the soil bed.

WEARING SURFACES

Functions of Wearing Surface. The office of the wearing surface is to protect the foundation from the wear of the traffic and the effects of surface water, and to support the weight of the traffic and transmit it to the foundation. To render efficient service to the traffic, it must furnish a comparatively smooth unyielding surface that affords good foothold for draft animals and good adhesion for motor vehicles, and on which the resistance to traction will be a minimum. To fill its office satisfactorily the material of which it is composed must possess strength to resist crushing and abrasion, and its fabric must be practically impervious. To render economical service, it must possess the power of resisting the action of the destroying agencies for a reasonable length of time before it becomes unfit for use. For this purpose it must possess the resisting qualities previously stated, and it must also possess a certain thickness; this thickness will depend upon the character of the material employed and its rate of wear under the given traffic and atmospheric conditions. Economy is not promoted by using a thick wearing surface, as under heavy traffic it will be so worn in a few years as to be unserviceable, and under light traffic it will be decomposed before it is worn out. In either case it must be removed and the portion so removed is waste; therefore, only such thickness as will give efficient service during a few years should be adopted.

Thickness. The measure for the economical thickness of any type of wearing surface is that the annual interest charge on the first cost plus the annual depreciation shall be a minimum. To apply this measure it is necessary to know the amount of traffic and the loss of thickness due to wear.

Classification of Wearing Surfaces. The wearing surfaces most commonly employed for roads and streets are composed of: (1) gravel, broken stone, furnace slag, and similar granular materials bound with colloidal cement formed by the action of water on the plastic elements of rock and clay; (2) broken stone, gravel, and sand bound with: (a) bituminous cement; (b) hydraulic cement; (3) stone blocks; (4) brick; (5) wood blocks.

In type (1), a certain amount of moisture is essential to successful binding. When this is lacking, as in the summer season, the binding material becomes dry and brittle, and the fragments at the surface are displaced by the action of the traffic; an excess of moisture destroys the binding power; and the surface is quickly broken up by the traffic.

Wearing surfaces of type (2a) are usually limited in life not merely by the wear of traffic, but by the fact that all bitumens slowly alter in chemical composition when exposed to atmospheric action, and in time become brittle. Type (2b) is subject to cracking under expansion and contracting, due to changes of temperature, and is liable to wear unevenly owing to irregularity in mixing and the segregation of the ingredients while the concrete is being put in place. When a defective spot begins to wear, it extends very rapidly under the abrasive action of the traffic.

The materials of types (3) and (4) seldom rot or disintegrate and, when the pavement is well constructed, are eminently enduring and generally render satisfactory service. Since the use of creosote and other preservatives has increased the service life of wood blocks, type (5), by lessening their tendency to decay, they have come into extensive use for street paving.

Gravel Roads

Gravel. Gravel consists of smooth and somewhat rounded stones, varying in size from small grains to pebbles 4 or more inches in diameter. It is found mixed with sand, on the banks and in the beds of rivers; and in deposits on the land, mixed with clay and other mineral substances, such as limestone and oxide of iron, from which it derives a distinctive name. Gravel of the latter class is called cementatious and when suitably prepared cements together, forming a very satisfactory roadway for light traffic, producing but little dust in dry weather and costing little to maintain.

Preparation of Gravel. Gravel is best prepared for use by screening into three grades: grade (1), containing the stones retained by a $1\frac{1}{2}$-inch mesh screen and passing a $2\frac{1}{2}$-inch mesh; grade (2), containing the stones retained by a $\frac{1}{4}$-inch mesh and passing a $1\frac{1}{2}$-inch mesh; grade (3), containing all the material passing the $\frac{1}{4}$-inch mesh. The voids in grade (1) are determined, and enough of grade (3) added slightly more than to fill them; the two are intimately and evenly mixed and the mixture is used for the first or lower course. The voids in grade (2) are determined and a sufficient quantity of grade (3) added to fill them; the two are mixed and used for the top course. The mixture should be combined very evenly so that the fine material is mixed uniformly with the coarse; and in spreading the mixture, care should be taken to avoid separating it or allowing the fine material to settle to the bottom.

If the gravel is deficient in binding material, the latter may be added in the form of clay, loam, limestone screenings, shale, or marl, the amount added ranging from 10 to 15 per cent. An excess (20 per cent) of clay causes the gravel to pack quickly and to present a good appearance under the rolling; but in dry weather the road will ravel, become defective and dusty, and in wet weather it will be muddy. Clean smooth gravel will not consolidate without a binder and, unless this is of very good quality, a road made with it will prove unsatisfactory.

Laying the Gravel. On the natural-soil bed properly graded and compacted, the prepared gravel is spread uniformly to the depth desired—usually 6 inches. Then it is compacted by rolling with a steam roller, after which it is moistened by sprinkling with water, and the rolling is repeated. The sprinkling and rolling are repeated as often as may be required, until the stones cease to rise or creep in front of the roller. The second course then is spread to a depth of about 4 inches, rolled, sprinkled, and again rolled in the same manner and to the same extent as the first course. After this, a thin coat of the fine screenings is spread over the surface and the traffic is admitted.

If, during the rolling, the first course appears to be deficient in binding material, more may be added by spreading a thin layer of the fine material over the surface of the course, sprinkling and rolling, as above described.

If, during the rolling of the top course, any stones larger than $1\frac{1}{2}$ inches appear, they must be removed.

Gravel shrinks in rolling about 20 per cent of its loose depth; therefore, to obtain a thickness of 8 inches when compacted, the loose material should have a depth of about 10 inches. The thickness of the gravel coating varies according to the nature of the roadbed, a thicker layer being necessary on impermeable soil than on a well-drained soil.

The pebbles in a gravel road are imbedded in a paste and can be displaced easily. It is for this reason, among others, that such roads are subject to internal destruction.

The binding power of clay depends in a large measure upon the state of the weather. During rainy periods a gravel road becomes soft and muddy, while in very dry weather the clay will contract and crack, thus releasing the pebbles, and causing a loose surface. The most favorable conditions are obtained in moderately damp or dry weather, during which a gravel road offers several advantages for light traffic, the character of the drainage, etc., largely determining durability, cost, maintenance, etc.

Repair. Gravel roads constructed as above described will need only small repairs for some years, but daily attention is required in making these. A garden rake should be kept at hand to draw any loose gravel into the wheel tracks, and for filling any depressions that may occur.

In making repairs, it is best to apply a small quantity of gravel at a time, unless it is a spot which actually has cut through. Two inches of gravel at once is more profitable than a larger amount. Where a thick coating is applied at once it does not all pack, and if, after the surface is solid, a cut be made, loose gravel will be found; this holds water and makes the road heave and become spouty under the action of frost. It will cost no more to apply 6 inches of gravel at three different times than to do it at once.

At every $\frac{1}{8}$ mile a few cubic yards of gravel should be stored to be used in filling depressions and ruts as fast as they appear, and there should be at least one laborer to every 5 miles of road.

Broken-Stone Roads

Methods of Construction. Broken-stone roads are formed in several different ways. For example, the road may be formed by placing one or two layers of stone broken into small fragments upon: (1) the natural soil; (2) a foundation composed of large stone set by hand upon the natural soil; or (3) a foundation layer of cement

Fig. 61.　Views Showing Method of Building Telford Roads for Pennsylvania State Highway Department

concrete. The layers of broken stone are compacted by rolling with a heavy roller and the interstices, or spaces between the stones are filled with a binder composed either of stone dust; stone dust and clay; a grout of hydraulic or Portland cement; or a bituminous cement derived from either coal tar or asphalt, and used alone or mixed with sand. The broken stone forming the lower surface layer often is coated with a bituminous cement before placing it upon the foundation. This applies particularly when the upper layer is of bituminous cement. The broken stone also may be mixed with Portland cement and sand, forming a concrete, which is placed either upon the prepared natural soil or upon a concrete or broken-stone foundation.

The several methods for constructing broken-stone roads are distinguished by either a specific name or the name of the introducer. Thus, the types known as *Telford* and *Macadam* are named from Thomas Telford and John L. McAdam, Scottish engineers, who introduced them in England during the early part of the 19th Century, as an improvement of the method employed in the 18th Century by M. Tresaguet on the roads of France. Telford used a base of large stones, Fig. 61, upon which the small stone was placed. McAdam omitted the base contending that it was useless and injurious. Both constructors insisted on thorough drainage of the subsoil, but neither used a binder and rolling was unknown. The stones were left to be compacted by the traffic. The introduction of stone-crushing machinery and rollers as well as the practice (condemned by McAdam, but advocated by Mr. Edgeworth, an Irish landowner in his treatise on Road Building published in 1817) of filling the voids with a binder has caused material departures from the methods of the pioneers whose names are still but improperly applied.

The cement grouting was introduced in England by Sir John Macneil. The coating of the stone with coal tar was first practiced in England about 1840, and was called "tar-macadam". In recent times, to distinguish the several varieties of bituminous construction, several specific terms have been coined, as "bitulithic", "tarmac", "warrenite", "bituminous macadam", "asphalt macadam", etc. Since the use of bituminous binding has become extensive, the term "water-bound macadam" has come into use, to distinguish the earlier macadam type from the types recently introduced.

Quality of Stones. The materials used for broken-stone pavements of necessity must vary very much according to the locality. Owing to the cost of haulage, local stone generally must be used, especially if the traffic be only moderate. If, however, the traffic is heavy, it sometimes will be found better and more economical to obtain a superior material, even at a higher cost, than the local stone; and in cases where the traffic is very great, the best material that can be obtained is the most economical.

There are a number of qualities required in a stone to render efficient service. *Hardness* and *toughness,* to resist the effects of abrasion and impact. These two properties, while closely related, are not always coincident; some rocks, although extremely hard, yet are so brittle that they crush easily under impact. In others the cohesion between the component particles is so weak that they are worn quickly by abrasion. *Durability,* or power to resist the disintegrating influences of the weather and humus acids. The quality of durability depends chiefly upon the chemical stability of the minerals present. Physical defects and abrasion generally cause the destruction of the stone long before it is injured by chemical changes. *Capability of binding* into a compact mass. This quality is essential to stone used for water-bound macadam. The binding or cementing property is possessed to a greater or less extent by all rocks when in a state of disintegration. It is caused by the action of water upon the chemical constituents of the stone contained in the detritus—material worn off—produced by crushing the stone and by the friction of the fragments on each other while being compacted; its strength varies with the different species of rock, but it exists in some measure with them all, being greatest with limestone and least with gneiss.

The essential condition of the stone to produce this binding effect is that it be sound. No decayed stone retains the property of binding, though in some few cases, where the material contains iron oxides, it may, by the cementing property of the oxide, undergo a certain amount of binding.

A stone of good binding nature frequently will wear much better than one without, although it is not so hard. A limestone road well made and of good cross section will be more impervious than any other, owing to this cause, and will not disintegrate so soon in dry weather, owing partly to this and partly to the well-known quality which all limestone has of absorbing moisture from the atmosphere. Mere hardness without toughness is not of much use, as a stone may be very hard but so brittle as to be crushed to powder under a heavy load, while a stone not so hard but having a greater degree of toughness will be uninjured.

A stone for a road surface should be as little absorptive of moisture as possible in order that it may not suffer injury from the action of frost. Many limestones are objectionable on this account.

The stone used should be uniform in quality, otherwise it will wear unevenly, and depressions will appear where the softer material has been used. As the under parts of the road covering are not subject to the wear of traffic, and have only the weight of loads to sustain, it is not necessary that the stone of the lower layer be so hard or so tough as the stone for the surface, hence it is frequently possible by using an inferior stone for that portion of the work, to reduce greatly the cost of construction.

Testing the Rock. In order to ascertain the probable resistance of the different rocks to the destructive action of the traffic and weather, tests are made in the laboratory to determine the resistance to impact and abrasion, absorptive capacity, hardness, toughness, and specific gravity.

Abrasion. The test for abrasion is conducted in the Deval type of machine. It consists of two or more cast-iron cylinders mounted on a shaft so that the axis of each cylinder is inclined an angle of 30 degrees from the axis of rotation. The cylinders are charged with 11 pounds of the rock broken into fragments, ranging from $1\frac{1}{4}$ to $2\frac{1}{2}$ inches. The cylinders are then rotated at a uniform speed of 2,000 revolutions per hour for five hours, or until the automatic recorder shows 10,000 revolutions; the charge then is removed and placed on a sieve having meshes of $\frac{1}{16}$ inch. The material retained on the sieve is washed, dried, and weighed. The difference in weight between the weight of the charge and the residue larger than $\frac{1}{16}$ inch shows the loss by abrasion.

Impact and Toughness. The test for impact and toughness is made in a machine, consisting of an anvil, plunger, and hammer, mounted in vertical guides. The test piece is placed on the anvil; the hammer weighing 4.40 pounds is raised and allowed to fall a distance of one centimeter for the first blow and an increased fall of one centimeter for each succeeding blow, until the test piece fails. The number of blows required to destroy it is used to represent the toughness; 13 blows is considered to indicate low resistance, 13 to 19 medium, and above 19 high.

Hardness. The test for hardness is made on a Dory machine, which consists of a steel disk mounted so as to be revolved. The test pieces are cylinders cut from the rock by a core drill, and the ends ground level. Two pieces are used for a test; each is weighed, then placed in the guides of the machine with its face resting upon the grinding disk. The machine is revolved until 1000 revolutions have been made, and during the operation, quartz sand is fed onto the disk. The test piece is removed and weighed, and the hardness is determined from the formula

$$\text{Hardness} = 20 - \frac{W}{3}$$

in which W is loss in grams per 1000 revolutions. Rocks having a hardness less than 14 are considered soft; from 14 to 17 medium; and over 17 hard.

Water Absorption. The capacity of the stone to absorb water is determined by using a thoroughly dry sample of stone weighing about 12 grams. The sample is weighed in air, then immersed in water where it is weighed immediately; after 96 hours' immersion it is weighed again in the water. The absorptive capacity then is calculated by the formula

$$\text{Lb. water absorbed} = \frac{C-B}{A-B} \times 62.37 \text{ per cu. ft. of rock}$$

in which A is the weight in air; B is the weight in water immediately after immersion; C is the weight in water after immersion for 96 hours; and 62.37 is the normal weight in pounds of a cubic foot of water.

The durability of a stone used for roads is affected to a certain extent by its capability of absorbing water. In cold climates a low absorptive capacity is essential to resist the disintegrating effects of alternate freezing and thawing.

Specific Gravity. The specific gravity is determined either by weighing in a specific gravity balance or by weighing in air and water, and applying the formula

$$\text{Specific gravity} = \frac{W}{W - W_1}$$

in which W is the weight in air, and W_1 is the weight in water.

Specific gravity and porosity are closely related. The specific gravity varies with the density or compactness of the aggregation of the mineral grains forming the stone. The closer the grains the more compact the stone, and the less will be the amount of interstitial space and hence the less the porosity.

From the specific gravity the weight per ton or per cubic yard may be determined. A knowledge of the weight is useful in deciding between two otherwise good stones; the heavier will be the more expensive, due to increased cost of transportation. On a water-bound macadam road it is an advantage to have a detritus with a high specific gravity, as it will not be moved so easily by rain and wind as one of low specific gravity.

Cementing Quality. The cementing quality of the stone dust is determined by placing 500 grams of the rock, broken to pass a ½-inch mesh screen, in a ball mill, together with 90 cubic centimeters of water and 2 steel balls weighing 20 pounds. The mill and its charge are revolved for 2½ hours at a rate of 2000 revolutions per hour. The operation produces a stiff dough, of which 25 grams are placed in a metal die 25 millimeters in diameter, and subjected to a pressure of 132 kilograms per square centimeter, producing a cylindrical test piece. The test piece is dried in the air for 20 hours, after which it is heated in a hot-air oven for 4 hours at a temperature of 200° Fahrenheit and then cooled in a desiccator for 20 minutes. When cool it is tested in the impact testing machine in the same manner as the test for toughness, using a hammer weighing 1 kilogram and a fixed height of fall of 1 centimeter. Blows are struck until the test piece fails. The average of the number of blows on 5 test pieces is taken as the result of the test. A result of 10 is considered to indicate a low cementing quality; 10 to 25 is considered fair; 26 to 75 good; 76 to 100 very good; over 100 excellent.

Species of Stone. The rocks most extensively used for broken-stone roads are trap, granite, limestone, sandstone, boulders, or field-stone.

Trap rock is hard and tough and has good wearing and binding qualities. Granite is brittle and its cementing value is low. Limestone is deficient in hardness and toughness but possesses good binding qualities and for light traffic roads is eminently suitable. Sandstones are rocks made up of grains of sand cemented together by siliceous, ferruginous, calcareous, or argillaceous material; they are usually deficient in binding quality and resistance to abrasion. With a bituminous binder good results are obtained. Fieldstones are a mixture of the hardest parts of the granites, sandstones, limestones, etc., distributed by glacial action and which have resisted the disintegrating effects of the weather. Owing to their variable character and unequal hardness they wear irregularly and make a very rough road.

. **Shape and Size of Stone.** The shape and size of the fragments of stone affect the enduring qualities of the road. The nearer the fragments approach the cubical form with irregular jagged sides, the more satisfactory will be the results.

The size of the stone must be such as will not fracture or crush under the action of the roller during compaction nor become loosened under traffic. For the harder rocks the size varies from $1\frac{1}{2}$ to $2\frac{1}{2}$ inches and for the softer rocks from $2\frac{1}{2}$ to 4 inches. The sizes do not refer to the actual dimensions of the stone, but to the size of the hole in the screen; thus $1\frac{1}{2}$-inch stone is that which is retained by a $1\frac{1}{2}$-inch opening and passes through a 2- or $2\frac{1}{4}$-inch opening.

Thickness of the Broken Stone. The thickness of the broken stone is determined for the given conditions by the formula previously stated, and ranges from 4 to 16 inches. Where the thickness exceeds 6 inches, the excess may be composed of sand, gravel, fieldstone, ledge rock, or broken stone, as previously stated in the discussion on foundations; the choice depending on availability and cost. For use in the base all are equally effective.

Spreading the Stone. The method employed for laying the covering varies with the thickness. When the finished thickness is 4 inches all the stone to be used is laid in one course. When the finished thickness exceeds 4 inches the stone is laid in two or more courses; the top or wearing course, being composed of the best and most expensive stone, is made the least thickness compatible with good construction and maintenance. To provide for the shrinkage of the stone under the roller, the depth of the courses of loose stone should exceed the finished depth by from 25 to 30 per cent.

The stone is hauled upon the roadbed in vehicles of various types provided with broad-tired wheels. In some types of vehicle it is spread in layers as the vehicle is drawn along the roadbed; with others it is dumped in heaps and spread by hand with forks and brought to an even surface by raking, Fig. 62.

Compacting the Broken Stone. The stone is compacted by rolling with heavy rollers drawn by horses or propelled by steam or other power, Fig. 63. The steam roller is more effective than horse-

Fig. 62. View Showing Spreading of Lower Course of Macadam Road
Courtesy of United States Department of Agriculture

drawn rollers. The usual weights of steam rollers are 5, 10, and 15 tons; the 10-ton being the one generally used, although the weight of the roller should be selected in accordance with the bearing power of the natural soil. A roller having excessive weight may cause injury to the roadbed, by rolling it into undulations that will permit water to collect and consequently cause damage. A roadbed which will stand a heavy roller in dry weather may be injured by it during wet weather. For a weak roadbed it is well to use two rollers, one of light weight to form a crust and a narrow, heavy roller to compact it.

The roller should commence at one edge or border of the roadway, and move along that edge until within about 25 feet of one end of the spread stone; it then should cross over to the other edge and proceed along this edge to the beginning, crossing over and overlapping the strips previously rolled until the center of the road is reached. The rolling is continued in this manner until the stones cease to creep in front or sink under the roller. If, during the first passages of the roller, low spots appear, they should be filled to grade with stone of the same size as is in the course being rolled.

Fig. 63.   Compacting Broken Stone by Steam Roller
Courtesy of United States Department of Agriculture

After about two passages of the roller, the binder, consisting of the screenings from the stone being used for the course, is spread in a thin layer over the surface of the partly compacted stone and sprinkled with water, which washes it into the voids in the stone; the rolling then is continued, Fig. 64. The operation of applying the binder, sprinkling, and rolling is repeated until a wave of water and screenings rises in front of the roller. Each course is treated and rolled in the same manner. If the screenings from the rock that is being used are not suitable for binding, screenings from other rock, clay, sand, or loam are substituted.

An excess of binder and water will shorten the time required to consolidate the stone and produce the appearance of a good piece of work, but under traffic it will wear unevenly and go to pieces quickly.

Suppression of Dust on Macadam and Telford Roads. Since the introduction of mechanically, propelled vehicles, broken-stone roads constructed according to the principles of Telford and McAdam, have proven inadequate to the demands of the changed traffic.

The adhesion between the particles of stone is insufficient to react against the propulsive force exerted by the driving wheels,

Fig. 64.   Rolling and Sprinkling Second Course of Macadam Road to Complete Binding Process
Courtesy of United States Department of Agriculture

hence the stones are loosened, and although the rubber tires with which the motor vehicles are equipped produce little dust by attrition or wearing away, the vehicle moving at high speed creates a partial vacuum. The current of air which then rushes in to re-establish the equilibrium picks up the small particles of stone displaced and loosened by the thrust of the driving wheels and distributes them in the form of dust, which is very disagreeable to other users of the road and residents along it. The large stones that are loosened are thrown about and ground upon one another and thus increase the amount of fine material ready to be scattered as dust.

The frequent repetition of these actions causes the pavement to pit and disintegrate. The destructive effect is intensified, the greater the speed, and where the irregularities of the surfaces are such as to cause the wheels to leave it, there is produced a bounding motion that is continued for some distance and is particularly disastrous. Shearing of the road fabric is very severe on steep grades and curves due to the slipping of the driving wheels when the propulsive force is greater than the adhesion between the tire and the road surface. The damage arising from this is more extensive during wet weather and is intensified when the wheels are equipped with bars, chains, studs, and other anti-skidding devices.

The formation of dust and mud cannot be prevented absolutely, because all materials, by attrition and the disintegrating action of the elements, yield dust when dry and mud when wet. If the surface of a water-bound macadam road could be maintained in a moist condition, there would be no dust, but moistening with water even in cities, towns, and villages is expensive, and in rural districts the cost is prohibitive and the practice would be impossible, owing to the absence of water available for the purpose. Hence in dealing with existing road surfaces a remedy has been sought in more frequent cleansing and in the use of some substitute for water which would be cheap, effective, lasting, and easily applied. To meet this demand several "dust-laying" compositions have been placed on the market, and experiments have been made with some of these, but it has been demonstrated clearly that, with but few exceptions, they have a very temporary effect, and their application must be frequent and thorough.

Under the head of exceptions, that is, of the more or less permanent methods, are included the following: (1) the cementing of the surface stone by a bituminous cement or binder. When the binder is applied by the penetration method, the surface is described by the general term "bituminous-macadam"; and when it is desired to indicate the kind of binder, the descriptive names, "asphalt-macadam", "tar-macadam", etc., are used. When the binder is applied by the mixing method, the construction is called "bituminous-concrete", or specifically designated by the trade or patented name as, "bitulithic", "warrenite", "amiesite", "filbertine", "rock-asphalt", etc.; (2) binding the stone with hydraulic cement, the surface so formed being called "concrete-macadam", or "concrete pavement". These will be discussed later under their respective headings.

Turning to the details of the various temporary methods, we find the following:

(1) *Fresh Water.* This is the simplest remedy, but not always the most practicable nor the cheapest.

(2) *Sea Water.* This is a simple remedy but available only on the seacoast. The salts contained in sea water are highly antiseptic and deliquescent; a light sprinkling will suppress the dust for several hours. Its use, however, is objected to for the reason that it injures the varnish and running gear of vehicles, corrodes cast-iron street fittings, and when the road surface on which it has been used has dried the dust then produced, containing salt, injures food and other goods exposed to it. Moreover, after a few weeks' use the dust is converted into a pasty mud that adheres to the wheels and causes the surface of the road to be "picked up".

(3) *Deliquescent Salts.* The chief advantage of these salts is that their effect is more lasting than that of water. The salt used most extensively is calcium chloride obtained as a by-product in the manufacture of soda by the ammonia process. The salt may be applied either in solution or in the dry form. It takes up water rapidly and proves very efficient where the atmospheric moisture is sufficient to feed the salt. Glutrin, the commercial name for the waste sulphite liquor obtained in the manufacture of paper from wood pulp by the sulphite process, reduces the formation of dust, but the treatment must be repeated frequently. Waste molasses or "black strap" from sugar refineries mixed with milk of lime possesses good dust-suppressing qualities.

(4) *Coal-Tar Coating.* Refined coal tar applied either hot or cold in the form of a spray minimizes the production of dust, renders the surface waterproof, and reduces wear. The success attending its use depends upon the quality of the tar, the state of the weather, which must be clear and dry, the condition of the road surface, which must be dry and free from dust and dirt, and, in the case of hot application, that the tar is not overheated.

(5) *Solutions of Coal Tar and Petroleum.* Several patented preparations of coal tar are on the market. The principle of all is practically the same, namely, the solution of the tar or oil in water by a volatile agent, which on evaporation leaves a more or less insoluble coating on the road surface. The more favorably known of these preparations are "tarvia" and "westrumite".

(6) *Crude Petroleum and Residuum Oil.* Crude petroleum containing a large percentage of asphalt gives the best results. Petroleum having paraffin and naphtha as a base refuses to bind, and produces a greasy slime. The residuum oils obtained in the distillation of petroleum having asphaltum for a base have yielded good results in many cases.

Two methods are followed in applying the oil: (a) The surface of the road to be oiled is prepared by removing the dust with hand or power brooms. The oil, in the cold method, is applied by specially designed sprinkling wagons, at the rate of from one-third to one-half gallon per square yard. After being applied the oil is covered with sand or stone screenings and may or may not be rolled. The oil is applied once or twice a year according to whether the traffic is light or heavy. The surface of the road must be dry when the oil is applied.

(b) The oil is sprinkled over the surface and mixed with the dust. If the oil is merely sprinkled, the mixture of dust and oil made by the action of the traffic will become very sticky and will be removed in spots by adhering to the wheels. For the purpose of facilitating the handling and of securing a deeper penetration than is possible with cold oil, the oil is heated to a temperature of about 140° Fahrenheit and applied in the same manner as the cold oil.

(7) *Oil Tar and Creosote.* Oil tar is the residual liquid from the manufacture of carbureted water gas and oil gas. The tar used for road purposes is obtained by distilling the original tarry liquid to remove the light oil, naphthalene, and creosote. Various grades of tar are produced according to the temperature at which the distillation is stopped. The higher the temperature of distillation, the harder and more brittle the tar.

The oil tar either alone or mixed with creosote is applied in the same manner as coal tar.

Bituminous=Macadam

Features of Bituminous=Macadam. A bituminous-macadam wearing surface differs from the previously described water-bound broken-stone surface only in the kind of binder and the quality of the stone. The bituminous binder is prepared from asphalt, asphaltic oils, refined water-gas tars, refined coal tars, and combinations of refined tars and asphalts.

The bituminous binders adhere to comparatively porous and relatively soft stone, such as limestone, better than to the hard stones, such as trap and granite. Consequently, the stone used with the bituminous binders may be inferior in hardness and binding quality to that required for water-bound macadam.

Methods of Construction. The essentials necessary to the successful construction of a bituminous covering are: (1) the exclusion of both subsoil and surface water from the foundation; (2) a solid unyielding foundation; (3) a stone of suitable quality and size; (4) that the stone shall be entirely free from dust, otherwise the dust will interpose a thin film between the stone and the bituminous binder and prevent the latter from adhering to the stone; (5) if the stone is to be used hot, that it shall not be overheated; and if is to be used cold, that it shall be dry, for if wet or damp, the bituminous material will not adhere to it; (6) that the bituminous cement shall be of suitable quality; free from water, for which the stone has a greater affinity than for bitumen, and would thus prevent adhesion; free from ammoniacal liquor, which is apt to saponify some of the oily constitutents and thus render them capable of combining with water and therefore apt to be washed out; free from an excess of light oils and naphtha, which act as diluents and volatilize on the surface of the road, forming a skin that is not durable; free from an excess of free carbon, because it has no binding value and is liable to be converted into dust and mud.

Two general methods with various modifications in the minor details are employed for applying the bituminous binder to form the wearing surface, viz, the penetration method, and the mixing method.

Penetration Method. In this method, the stone is spread and packed slightly by rolling. The bituminous binder is then applied by one of the following ways: by hand from pouring pots; by a nozzle leading from a tank cart; or by a mechanical distributor using air pressure to discharge the material through nozzles that spread it in a finely divided stream or spray, Fig. 65. The binder is heated, usually by steam from the roller, but when hand pots are used, it is heated in kettles over fires. The quantity applied is about 1½ gallons per square yard. After the binder is distributed, it is covered with a light coating of stone dust, sand, or gravel, and the rolling is continued. In some cases, after the rolling is completed, another application of the binder is made at the rate of about one-half gallon per square yard; this is called a "paint coat" and is covered with a light sprinkling of stone screenings.

Fig. 65. Spreading Bituminous Binder by Pressure Nozzle, Penetration Method
Courtesy of Barrett Manufacturing Company, New York City

Mixing Method. In this method the stone to be used for the wearing surface, varying in size from ½ to 1¼ inches, is cleaned and dried, then mixed with a sufficient quantity of the binder to coat all the stones thoroughly. The mixing is performed by manual labor on a mixing board, Fig. 66, or by raking the stones through a bath of liquid binder, or by passing through a mechanical mixing machine, Fig. 67. The coated stones are spread upon the foundation in a layer having a thickness of about 3 inches and are covered with a light coating of stone screenings free from dust; then are compacted by rolling, Fig. 68. Wherever the binder flushes to the surface it is covered with screenings and rolled. When the rolling is completed, the surplus screenings are swept from the surface. The cleaned surface then is covered with a coat of the

Fig. 66. Hand Method of Mixing Stone and Binder
Courtesy of Barrett Manufacturing Company, New York City

Fig. 67. Machine Method of Mixing Stone and Binder
Courtesy of Barrett Manufacturing Company, New York City

Fig. 68. Rolling Bituminous Macadam Road Surface
Courtesy of Barrett Manufacturing Company, New York City

binder called a "seal coat", for the purpose of insuring the water-proofing and complete filling of the voids, Fig. 69. For this coat, about one-half gallon of binder is used per square yard of surface. Screenings again are spread and may or may not be rolled.

Advantages and Disadvantages of the Penetration Method. The advantage of the penetration method is the ease and rapidity with which it can be carried out, and the low cost for equipment and labor.

The disadvantages of the penetration method are: (1) the difficulty of obtaining an absolutely uniform distribution of the binder, thus producing "lean" and "fat" spots that will prove defective under traffic; (2) it is wasteful, in that it is necessary to use more binder than actually is required to coat the stones and bind them together;

Fig. 69. Spraying Seal Coat by Auto Truck, One-Half Gallon to the Yard
Courtesy of Barrett Manufacturing Company, New York City

(3) it is difficult and sometimes impossible to use a binder of sufficient original consistency to produce a satisfactory bond, owing to the bitumen setting too rapidly when applied to cold stone.

Advantages and Disadvantages of the Mixing Method. The advantages of the mixing method are: (1) the producing of a uniform fabric in which the cement is distributed uniformly and cements each individual stone; (2) that construction can be carried on in colder weather than is permissible with the penetration method. If hot stone is used, a bitumen can be employed of such original consistency as is required to sustain the traffic satisfactorily.

The disadvantage of the mixing method is the greater cost,

due (1) to the increased labor, and (2) to the more elaborate equipment and apparatus required.

Bitulithic. Bitulithic is composed of stone, ranging in size from 2 inches to $\frac{1}{800}$ of an inch, and dust, which are dried, heated, and mixed in predetermined proportions, so as to reduce the voids to about 10 per cent, and cemented by a hot bituminous cement manufactured from either coal tar, asphalt, or a combination of both. The cement is added in sufficient quantity not only to coat every particle and to fill all of the remaining voids but with enough surplus to result in a rubbery and slightly flexible condition of the mixture after compression.

The mixture is spread, while hot, to such depth as will give a thickness of 2 inches after compressing with a 10-ton roller. After rolling, a composition coating called a "flush coat" is spread over the surface; this being covered while sticky with hot stone chips which are rolled until cool. The purpose of the stone chips is to form a gritty surface to prevent slipping.

Amiesite. Amiesite is a patented preparation of crushed stone or gravel, coated with an asphaltic cement. It is laid in two courses and a surface finish. The first course, composed of stone ranging from $\frac{1}{2}$ inch to $1\frac{1}{2}$ inches, is spread to a depth of 3 inches, blocks or strips of wood being used to insure uniformity of depth, then rolled once. The second course is composed of stone $\frac{1}{2}$ inch and less, spread 1 inch deep, then rolled. The surface finish consists of screenings or sand, used in sufficient quantity to fill the voids.

Rock Asphalt. The rock asphalt most used in the United States is a sandstone containing from 7 to 10 per cent of asphalt. It is prepared for use by pulverizing and is used either hot or cold. It is spread upon the surface of the stone to a depth of about $1\frac{1}{2}$ inches and rolled with a steam roller; the rolling is repeated daily for several days, or until the asphalt becomes hard.

Definitions of Bituminous Materials. The most recently adopted definitions of the bituminous materials employed in road construction are:

Native Bitumen. Native bitumen is a mixture of native or pyrogenous hydrocarbons and their non-metallic derivatives, which may be gases, liquids, viscous liquids, or solids and which are soluble in carbon disulphide.

Artificial Bitumen. Artificial bitumen is produced by the destructive distillation of pyrobitumens and other substances of an organic nature; the bitumens so produced are commonly known as tars, the word tar being compounded with the name of the material which has been subjected to the process of destructive distillation, thus designating its origin, as, coal tar, oil tar, etc.

Bituminous. Bituminous refers to that which contains bitumen or constitutes a source of bitumen.

Emulsions. Emulsions are oily substances made mixable with water through the action of a saponifying agent or soap.

Fixed Carbon. Fixed carbon is the organic matter of the residual coke obtained upon burning hydrocarbon products in a covered vessel in the absence of free oxygen.

Fluxes. Fluxes are fluid oils and tars which are incorporated with asphalt and semi-solid or solid oil and tar residuums for the purpose of reducing or softening their consistency.

Residuums, Residual Petroleum, or Residual Oils. These are heavy viscous residues produced by the evaporation or distillation of crude petroleums until at least all of the burning oils have been removed.

Bituminous Cement. The bituminous cements or binders are prepared from (1) coal-, oil-, and water-gas tars; (2) asphaltic petroleums; (3) asphalt; and (4) combinations of asphalt and the residues of distillation from asphaltic petroleums.

Coal-Tar Binder. Coal-tar binder is the residue obtained by the distillation of the crude tar produced in the manufacture of illuminating gas and the manufacture of coke for metallurgical purposes; the required consistency is obtained by removing part or all of the contained oils. Owing to the difference in the temperatures employed in the two producing processes, the constituents of the tar, while identical in their characteristics, differ in their amount; the most marked difference being in the free carbon content, of which coke-oven tar has the least.

Oil-Gas Tar. Oil-gas tar is produced in the manufacture of gas from oil. The tarry residue is rather varied in character and is prepared for use by distillation; it usually contains a large amount of free carbon.

Water-Gas Tar. Water-gas tar is produced in the manufacture of carbureted water gas for illuminating purposes and results from the petroleum product employed for carbureting. It is a thin oily liquid containing a large percentage of water. It is prepared for road use by mechanical dehydration and distillation. It has a strong gassy odor when applied to the road, but this disappears in a few days.

Asphaltic Petroleums. Asphaltic petroleums are native petroleums which yield asphalts upon reduction. They are used in the crude state, or after the illuminating and other oil constituents have been removed by cracking or blowing.

Asphalt. Asphalt is the name by which the native semi-solid and solid bitumen is known. Asphalt is the most permanent type of binder and has been used for many years in the construction of sheet-asphalt pavements. The semi-solid or tarry varieties are called "malthas" and are the ones generally employed as a binder for broken-stone roads; the solid variety is used almost exclusively for street pavements. Rock asphalt or bituminous rock is the name given to a great variety of sandstone rocks more or less saturated or cemented by maltha or hard asphalt. In referring to rock asphalts it is customary to add the name of the locality where they are found, as "Kentucky rock asphalt", "Italian rock asphalt", etc. The semi-solid varieties are used in the natural state or after the water and volatile hydrocarbons have been removed by heating. The hard varieties are prepared by softening with a suitable flux.

Combinations of Asphalt and Distillation Residues. Combinations of asphalt and the residue from the distillation of tars and petroleums are made by adding either a refined maltha or a pulverized solid asphalt; the mixing being accomplished by the injection of compressed air through suitably formed nozzles.

Tests for Bituminous Materials. The bituminous materials are subjected to certain tests for the purpose of ascertaining their chemical and physical properties. The results of the tests are used in specifications to secure the furnishing of the desired quality of material and to control the processes of manufacture; also to form a record by which the behavior of the materials under similar and dissimilar service conditions can be compared. The complex character of the materials requires a suitably equipped laboratory for the application of the tests.

The tests are determinations of (1) amount of water contained; (2) materials soluble in water; (3) homogeneity at a temperature of 77° Fahrenheit; (4) specific gravity; (5) consistency or viscosity measured by a standard penetration machine; (6) ductility, or the distance the material can be drawn out before breaking; (7) toughness, or resistance to fracture under blows in an impact machine; (8) melting or softening point, measured by a thermometer while the temperature is raised through a water or oil bath; (9) distillation—the products yielded at different temperatures during continuous distillation in a suitable flask or retort are caught and weighed; (10) amount of free carbon—a sample is dissolved in carbon bisulphide, the solution filtered, and the insoluble residue weighed; (11) amount of fixed carbon—a sample is placed in a platinum crucible and heated until the emission of flame and smoke ceases, then is allowed to cool and the residue is weighed; and the difference between the weight of the sample and the residue is the fixed carbon; (12) paraffine—the presence of paraffine is determined by treating a sample with absolute ether, freezing the mixture, filtering the precipitate, evaporating and weighing the residue; (13) amount of bitumen contained—found by weighing a sample of the dried material, by adding carbon bisulphide to dissolve the bitumen, and by drying and weighing the residue after the extraction is completed. The loss is the amount of bitumen soluble in carbon bisulphide. A sample also is treated with naphtha and the character of the residue is noted as to whether it is sticky or oily.

Concrete Pavements

Construction Methods. Several methods with many variations are employed for the construction of concrete pavements. The principal ones are: (1) grouting method, the construction being commonly called "concrete macadam"; (2) mixing method; (3) block or cube method.

Grouting Method. In this method the stone is spread upon the foundation and lightly rolled, after which a mixture of one part of cement and three parts of sand in the dry state is spread over the stone and swept into the interstices by brooms, then sprinkled with water and rolled; more cement and sand are spread, sprinkled, and rolled; the operation is repeated until the interstices are filled.

A variation of this method, known as the Hassam paving, is made by spreading the stone, ranging in size from $1\frac{1}{4}$ inches to $2\frac{1}{2}$ inches, and rolling it to a thickness of 4 inches, then filling the interstices with a grout composed of one part cement and three parts of sand mixed with water in a mixing machine, from which it flows over the surface, the machines being drawn along the roadway for this purpose; rolling is proceeded with at the same time and sufficient grout is applied to fill the interstices. On the foundation so prepared a wearing surface is formed; the stone is spread in the same manner as in the first course; the grout, composed of one part cement and two parts sand, mixed with sufficient water to make it very fluid, is applied by flowing over the surface of the compacted stone. The surface is finished by applying a thick grout composed of one part each cement, sand, and pea-sized trap rock.

Mixing Method. In this method the ingredients are combined into concrete by either hand or machine mixing; the concrete is deposited in place in one or two courses, the former being called "one-course" pavement and the latter "two-course" pavement. In the one-course method, the concrete mixed in the proportions of one part cement, one and one-half parts sand, and three parts stone is spread upon the prepared natural soil foundation and tamped to a thickness of about 6 inches.

In the two-course work the concrete mixed in the proportions of one part cement, two and one-half parts sand, and five parts stone is spread upon the prepared natural-soil foundation and compacted by rolling or tamping to the required thickness. On its surface, and before the cement has set, the wearing surface of about 2 inches in thickness is placed and tamped to the required contour. The mixtures used for the wearing surface vary, being composed of sand and cement, or of sand, cement, and small broken stone. The wearing surface of the Blome pavement is composed of one part cement and one and one-half parts of aggregate, which is made up of 50 per cent $\frac{1}{4}$-inch, 30 per cent $\frac{1}{8}$-inch, and 20 per cent $\frac{1}{10}$-inch granite screenings. The surface is formed into $4\frac{1}{2}$-inch by 9-inch blocks by cutting grooves $\frac{1}{2}$ inch wide and $\frac{1}{4}$ inch deep by means of special tools.

Materials. The materials used in the construction of concrete pavements should be selected with care. The stone should be a hard tough rock, free from dust and dirt, and graded so as to reduce voids to the minimum. The sand should be free from loam, clay, vegetable and organic matter, and should grade from coarse to fine. The cement should be of a quality to meet the standard tests. The water should be clean and free from organic matter, alkalies, and acids. Rapid drying of the concrete should be prevented by covering it with a canvas which is kept moistened with water for several hours; after its removal the surface should be covered with sand or earth which is to be kept moist for a period of two weeks. Improperly mixed or constructed concrete pavement will wear unevenly, crack, and rapidly become very defective.

Expansion Joints. To provide for the expansion and contraction of the concrete under changes of temperature, expansion joints are formed at intervals ranging from 15 to 50 feet. The edges of the joints are protected from injury by angle irons, and the space between them, about ½ inch, is filled with a bituminous cement which extends the full depth of the concrete. When the concrete is laid between curbs longitudinal joints from ½ inch to 1½ inches wide, filled with bituminous cement, are formed along the curb.

Reinforced=Concrete Pavement. Concrete pavements reinforced with steel in the form of woven-wire, Fig. 70, expanded metal, and round bars are constructed in two courses, the reinforcement being placed between the foundation course and the wearing surface.

Concrete with Bituminous Surface. In this type the surface of the concrete pavement, constructed by either the grouting or mixing method, is covered with a bituminous cement made from either asphalt, coal tar, or a mixture of both.

Block or Cube Pavement. In this type of pavement, blocks or cubes of concrete are molded in a machine or cast in molds. The blocks are stacked and allowed to season for three months, during which time they are wet twice a day. They are laid by hand on a sand cushion spread upon the foundation, then are brought to a firm bearing and uniform surface by rolling with a light roller. The surface is covered with a layer of sand or sandy loam which is broomed and flushed by water into the joints and the rolling is repeated; after which the surface is covered with a layer of sand, and the traffic then admitted.

A variation from the methods described is made in the patented pavement "rocmac". This is composed of broken stone cemented by silicate of lime, obtained by mixing powdered carbonate of lime with a solution of silicate of soda and sugar. The silicate of lime mortar is spread upon the foundation to a depth of about 2 inches, over which the broken stone is distributed to such a depth as will give, when compacted, a depth of about 4 inches. It is rolled and sprinkled with water until the mortar flushes to the surface, and

Fig. 70. Laying Reinforced Concrete Road. Woven Wire Fabric in Foreground Ready to Be Placed between Upper and Lower Coat
Courtesy of Municipal Engineering and Contracting Company, Chicago

then is covered with a layer of stone screenings and finally opened to traffic.

MAINTENANCE AND IMPROVEMENT OF ROADS

Repair and Maintenance of Broken=Stone Roads. These terms frequently but erroneously are used interchangeably. Repair means the restoring of a surface so badly worn that it cannot be maintained in good condition. A well-maintained road should not require repairs for a considerable length of time. The maintenance of a road is the keeping of it, as nearly as practicable, in the same condition as it was when constructed.

Good maintenance comprises: (1) constant daily attention to repair the ravages of traffic and the elements; (2) cleansing to remove the detritus caused by wear, the horse droppings, and other refuse; and (3) application of water or other dust layer.

When the surface of a water-bound broken-stone road requires to be renewed, it is loosened and broken up by scarifying, the new stone spread, rolled, watered, and bound in the same manner as in new construction; or the old surface is cleansed from dust and other matter by sweeping and washing, and the new stone spread upon it, compacted and finished as in new construction.

The resurfacing of water-bound roads with a bituminous construction is becoming common. The methods employed are the same as heretofore described under "bituminous macadam".

Systems of Maintenance. Several systems for maintaining roads are in use, the one yielding the best results being that which provides for the continuous employment of skilled workmen. The men so employed become familiar with the peculiarities of the sections in their charge and with the best way to deal with them. Efficient maintenance requires that the surfaces be kept smooth so that surface water may flow away rapidly and that the injury caused by traffic on uneven surfaces may be avoided; that incipient ruts, hollows, and depressions be eliminated by cutting out the area involved in the form of a square or rectangle and filling with new material; that dust and horse droppings be removed; that loose stones be removed; that gutters be clear so the rain water may be removed quickly; that ditches and culverts be cleaned out in advance of the spring and fall rains; that weeds and grass be removed from the earth shoulders, and that these and the dust sweepings be not left on the sides of the road to be redistributed, but be removed immediately and disposed of in such manner as will not cause injury; that bridges be examined and repaired at least twice a year.

Improvement of Existing Roads. The improvement of existing roads may be divided into three branches: (1) rectification of alignment; (2) drainage; (3) improvement of the surface.

The first of these consists in the application of the principles which have been laid down for the location, etc., of new roads and will include straightening the course by eliminating unnecessary curves and bends; improving the grade either by avoiding or cutting down hills and by embanking valleys; increasing the width where necessary, and rendering it uniform throughout.

The second, or drainage, consists in applying the principles laid down for the drainage of new roads, and in constructing the works necessary to give them effect.

The third, or improvement of the surface, consists in improving the surface by any of the methods previously described and that the funds available will permit.

Value of Improvement. The improvement of roads is chiefly an economical question relating to the waste of effort and to the saving of expenditure. Good roads reduce the resistance to locomotion, and this means reduction of the effort required to move a given load. Any effort costs something, and so the smallest effort costs the least, and therefore the smoothest road saves the most money for everyone who traverses it with a vehicle.

Cost of Improvement. Before undertaking any improvement generally it is required to know the cost of the proposed improvement and the benefits it will produce. In the improvement of roads the amount of money that may be expended profitably for any proposed improvement may be calculated with sufficient accuracy by obtaining first the following data: (1) the quantity and quality of the traffic using the road; (2) the cost of haulage; (3) plan and profile of the road; and (4) character and cost of the proposed improvements. From the data ascertain the total annual traffic and the total annual cost of hauling it. Next, calculate the annual cost of hauling the given tonnage over the road when improved. Then the difference between the two costs will represent the annual interest on the sum that may be expended in making the improvement. For example, if the annual cost of haulage over the existing road is $10,000 and the cost for hauling the same tonnage over the improved road will be $7000, the difference, $3000, with money at 6 per cent per annum, represents the sum of $50,000 that logically may be appropriated to carry out the improvement.

Traffic Census. The direction, character, and amount of traffic using a road is obtained by direct observation during different seasons of the year. As a preliminary to observing the traffic it is usual to determine the weight of the vehicles; this is done by weighing typical vehicles and by establishing an average weight for each type. The traffic is classified according to the motive power—as horse-drawn vehicles and motor vehicles. Each of these classes is

TABLE XI

Traffic Census

Average Hours per Day..................; for...........Days

Taken at

By

		EMPTY VEHICLES		LOADED VEHICLES	
		Nov. to March	Aug. to Oct.	Nov. to March	Aug. to Oct.
Horse	Pleasure				
	Commercial { Light Medium Heavy				
Motor	Pleasure { Motorcycles Runabouts Touring cars				
	Commercial { Light Medium Heavy				

divided into pleasure and commercial traffic, the latter class being subdivided into loaded and non-loaded vehicles. The number of horses to a vehicle in horse-drawn traffic and the speed of motor vehicles may be noted. A summary of data is suggested in Table XI.

The observations are made from 6 a. m. to 6 p. m., during a period of seven days each month, with occasional observations, from 6 p. m. to 6 a. m. or for the entire 24 hours if the amount of traffic requires it.

The weight of the traffic is ascertained by multiplying the number of each kind of vehicle by the average weight established for that type.

LAYING TRINIDAD SHEET ASPHALT PAVING IN FIFTH AVENUE, NEW YORK CITY

HIGHWAY CONSTRUCTION

PART II

CITY STREETS AND HIGHWAYS

The first work requiring the skill of the engineer is the laying out of town sites properly, especially with reference to the future requirements of a large city, where any such possibility exists. Few if any of our large cities were so planned. The same principles, to a limited extent, are applicable to all towns or cities. The topography of the site should be studied carefully, and the street lines adapted to it. These lines should be laid out systematically, with a view to convenience and comfort, and also with reference to economy of construction, future sanitary improvements, grades, and drainage.

Arrangement of City Streets. Generally, the best method of laying out streets is in straight lines, with frequent and regular intersecting streets, especially for the business parts of a city. When there is some centrally located structure, such as a courthouse, city hall, market, or other prominent building, it is very desirable to have several diagonal streets leading thereto. In the residence portions of cities, especially if on hilly ground, curves may replace straight lines with advantage, by affording better grades at less cost of grading, and by improving property through avoiding heavy embankments or cuttings.

Width of Streets. The width of streets should be proportioned to the character of the traffic that will use them. No rule can be laid down by which to determine the best width of streets; but it may be said safely that a street which is likely to become a commercial thoroughfare should have a width of not less than 120 feet between the building lines—the carriage-way 80 feet wide, and the sidewalks each 20 feet wide.

In streets occupied entirely by residences a carriage-way 32 feet wide will be ample, but the width between the building lines may be as great as desired. The sidewalks may be any amount over 10 feet which fancy dictates. Whatever width is adopted for them, not more of it than 8 feet need be paved, the remainder being occupied with grass and trees.

Street Grades. The grades of city streets depend upon the topography of the site. The necessity of avoiding deep cuttings or high embankments which seriously would affect the value of adjoining property for building purposes, often demands steeper grades than are permissible on country roads. Many cities have paved streets on 20 per cent grades. In establishing grades through unimproved property, they usually may be laid with reference to securing the most desirable percentage within a proper limit of cost. But

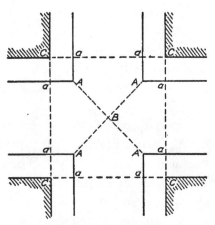

Fig. 71. Diagram Showing Arrangement of Grades at Street Intersections

when improvements already have been made and have been located with reference to the natural surface of the ground, the matter of giving a desirable grade without injury to adjoining property frequently is one of extreme difficulty. In such cases it becomes a question of how far individual interests shall be sacrificed to the general good. There are, however, certain conditions which it is important to bear in mind: (1) That the longitudinal crown level should be sustained uniformly from street to street intersection, whenever practicable. (2) That the grade should be sufficient to drain the surface. (3) That the crown levels at all intersections should be extended transversely, to avoid forming a depression at the junction.

Arrangements of Grades at Street Intersections. The best arrangement for intersections of streets when either or both have much inclination is a matter which requires careful consideration and upon which much diversity of opinion exists. No hard or fast rule can be laid down; each will require special adjustment. The best and simplest method is to make level the rectangular space *aaaaaaaa*, Fig. 71, with a rise of one-half inch in 10 feet from *AAAA* to *B*, placing gulleys at *AAAA* and the catch basins at *ccc*. When this method is not practicable, adopt such a grade (but one not exceeding 2½ per cent) that the rectangle *AAAA* shall appear to be nearly level; but to secure this it must have actually a considerable dip in the direction of the slope of the street. If steep grades are continued across intersections, they introduce side slopes in the streets thus crossed, which are troublesome, if not dangerous, to vehicles turning the corners, especially the upper ones. Such intersections

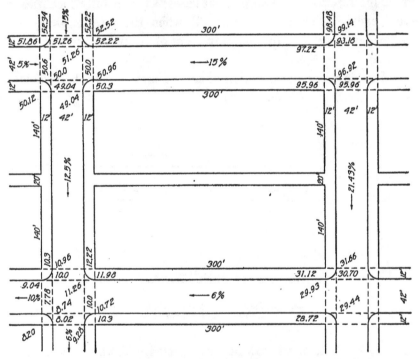

Fig. 72.   Diagram Showing Arrangement of Intersections for Steep Grades in
Duluth, Minnesota

are especially objectionable in rainy weather. The storm water will fall to the lowest point, concentrating a large quantity of water at two receiving basins, which, with a broken grade, could be divided among four or more basins.

Fig. 72 shows the arrangement of intersections in steep grades adapted for the streets of Duluth, Minnesota. From this it will be seen that at these intersections the grades are flattened to 3 per cent for the width of the roadway of the intersecting streets, and that the grade of the curbs is flattened to 8 per cent for the width of the

intersecting sidewalks.   Grades of less amount on roadway or sidewalk are continuous.   The elevation of block corners is found by
adding together the curb elevations at the faces of the block corners,
and $2\frac{1}{2}$ per cent of the sum of the widths of the two sidewalks at the
corner, and dividing the whole by two.   This gives an elevation
equal to the average elevation of the curbs at the corners, plus an
average rise of $2\frac{1}{2}$ per cent across the width of the sidewalk.

"Accommodation summits" have to be introduced between
street intersections in two general cases: (1) in hilly localities, to
avoid excessive excavation; and (2) when the intersecting streets
are level or nearly so, for the purpose of obtaining the fall necessary
for surface drainage.

The elevation and location of such a summit may be calculated
as follows: Let A, Fig. 73, be the elevation of the highest corner;

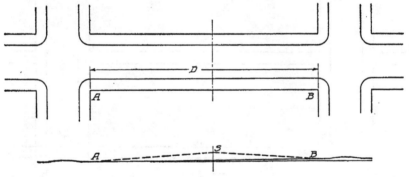

Fig. 73.   Diagrams for Calculating "Accommodation Summits" between
Street Intersections

B, the elevation of the lowest corner; D, the distance from corner
to corner; and R, the rate of the accommodation grade.   The
elevation of the summit is equal to

$$\frac{D \times R + A + B}{2}$$

The distance from A or B is found by subtracting the elevation of
either A or B from this quotient, and dividing the result by the rate
of grade.   Or the summit may be located mechanically by specially
prepared scales.   Prepare two scales divided to correspond to the
rate of grade; that is, if the rate of grade be 1 foot per 100 feet, then
one division of the scale should equal 100 feet on the map scale.

These divisions may be subdivided into tenths. One scale should read from right to left, and one from left to right.

To use the scales, place them on the map so that their figures correspond with the corner elevations; then, as the scales read in opposite directions, there is of course some point at which the opposite readings will be the same. This point is the location of the summit, and the figures read off the scale give its elevation. If the difference in elevation of the corners is such as not to require an intermediate summit for drainage, it will be apparent as soon as the scales are placed in position.

When an accommodation summit is employed, it should be formed by joining the two straight grade lines by a vertical curve, as described in Part I. The curve should be used both in the crown of the street and in the curb and footpath.

Where the grade is level between intersections, sufficient fall for surface drainage may be secured without the aid of accommodation summits, by arranging the grades as shown in Fig. 74. The curb is

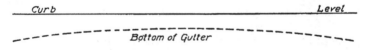

Fig. 74. Diagram Showing Arrangement of Grades to Avoid "Accommodation Summits"

set level between the corners; a summit is formed in the gutter; and receiving basins are placed at each corner.

Transverse Grade. In its transverse grade the street should be level; that is, the curbs on opposite sides should be at the same level,

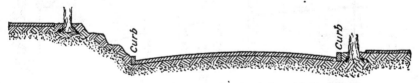

Fig. 75. Street with Unequal Transverse Grade but with Level Street

and the street crown rise equally from each side to the center. But in hillside streets this condition cannot be fulfilled always, and opposite sides of the street may differ as much as 5 feet. In such cases the engineer will have to use his discretion as to whether he shall adopt a straight slope inclining to the lower side, thus draining the whole street by the lower gutter, or adopt the three-curb method and sod the slope of the higher side.

In the improvement of old streets with the sides at different levels, much difficulty will be met, especially where shade trees have

Fig. 76. Street with Unequal Transverse Grade Inclined so as to Drain by Lower Gutter

to be spared. In such cases, recognized methods have to be abandoned, and the engineer will have to adopt methods of overcoming the difficulties in accordance with the conditions and necessities of each particular case. Figs. 75, 76, and 77 illustrate several typical

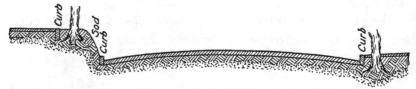

Fig. 77. Street with Unequal Transverse Grade with Three Curbs and Higher Slope Sodded

arrangements in the cases of streets where the opposite sides are at different levels.

Transverse Contour or Crown. The reason for crowning a pavement—i. e., making the center higher than the sides—is to provide for the rapid drainage of the surface. The most suitable form for the crown is the parabolic curve, which may be started at the curb line, or at the edge of the gutter adjoining the carriage-way,

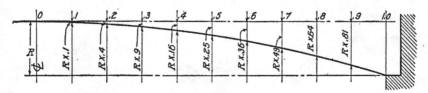

Fig. 78. Method of Obtaining Transverse Contour or Crown of a Road

about one foot from the curb. Fig. 78 shows this form, which is obtained by dividing the abscissa, or width from the center of the street to the gutter, into ten equal parts, and by dropping perpendiculars at each of these divisions, the lengths of which are determined by multiplying the rise at the center by the square of the

TABLE XII

Rise of Pavement Center above Gutter for Different Paving Materials

PAVING MATERIAL	PROPORTIONS OF RISE AT CENTER TO WIDTH OF CARRIAGE-WAY
Wood blocks	1 : 100
Stone blocks	1 : 80
Brick	1 : 80
Asphalt	1 : 80

successive values of the abscissas. The amounts thus obtained can be added to the rod readings; and the stakes, set at the proper distance across the street, with their tops at this level, will give the required curve.

The amount of transverse rise, or the height of the center above the gutters, varies with the different paving materials; smooth pavements requiring the least, and rough ones and earth the greatest. rise. The rise is generally stated in a proportion of the width of the carriage-way. The most suitable proportions are shown in Table XII.

Drainage of Streets. *Sub-Foundation Drainage.* The sub-foundation drainage of streets cannot be effected by transverse drains, because of their liability to disturbance by the introduction of gas, water, and other pipes.

Longitudinal drains must be depended upon entirely; they may be constructed of the same materials and in the same manner as road drains. The number of these longitudinal drains must depend upon the character of the soil. If the soil is moderately retentive, a single row of tiles or a hollow invert placed under the sewer in the center of the street generally will be sufficient; or two rows of tiles may be employed, one placed outside each curb line. If, on the other hand, the soil is exceedingly wet and the street very wide, four or more lines may be employed. These drains may be permitted to discharge into the sewers of the transverse streets.

Surface Drainage. The removal of water falling on the street surface is provided for by collecting it in the gutters, from which it is discharged into the sewers or other channels by means of catch basins placed at all street intersections and dips in the street grades.

Gutters. The gutters must be of sufficient depth to retain all the water which reaches them and prevent its overflowing on the footpath. The depth should never be less than 6 inches, and very rarely need be more than 10 inches.

Catch Basins. Catch basins are of various forms, usually circular or rectangular, built of brick masonry coated with a plaster of Portland cement. Whichever form is adopted, they should fulfill the following conditions:

(1) The inlet and outlet should have sufficient capacity to receive and discharge all water reaching the basin.

(2) The basins should have sufficient capacity below the outlet to retain all sand and road detritus, and prevent its being carried into the sewer.

(3) They should be trapped so as to prevent the escape of sewer gas. (This requirement frequently is omitted, to the detriment of the health of the people.)

(4) They should be constructed so that the pit can be cleaned out easily.

(5) The inlet should be constructed so as not to be choked easily by leaves or débris.

(6) They must offer the least possible obstruction to traffic.

(7) The pipe connecting the basin to the sewer should be freed easily of any obstruction.

The bottoms of the basins should be 6 or 8 feet below the street level; and the water level in them should be from 3 to 4 feet lower than the street surface, as a protection against freezing.

The capacity and number of basins will depend upon the area of the surface which they drain.

In streets having level or light longitudinal grades, gullies may be formed along the line of the gutter at such intervals as may be found necessary.

.Catch basins usually are placed at the curb line. In several cities, the basin is placed in the center of the street, and connects to inlets placed at the curb line. This reduces the cost of construction and cleaning, and removes from the sidewalk the dirty operations of cleaning the basins.

Catch basins and gully pits require cleaning out at frequent intervals; otherwise the odor arising from the decomposing matter contained in them will be very offensive. No rule can be laid down for the intervals at which the cleaning should be done, but they must be cleaned often enough to prevent the matter in them from putrefying. There is no uniformity of practice observed by cities in this matter; in some, the cleaning is done but once a year; in others, after every rain-storm; in still others, at intervals of three or four months; while in a few cities the basins are cleaned out once a month.

FOUNDATIONS

The stability, permanence, and maintenance of any pavement depend upon its foundation. If the foundation is weak, the surface soon will settle unequally, forming depressions and ruts. With a good foundation, the condition of the surface will depend upon the material employed for the pavement and upon the manner of laying it.

The essentials necessary to the forming of a good foundation are:

(1) The entire removal of all vegetable, perishable, and yielding matter. It is of no use to lay good material on a bad substratum.

(2) The drainage of the subsoil wherever necessary. A permanent foundation can be secured only by keeping the subsoil dry; for, where water is allowed to pass into and through it, its weak spots will be discovered quickly, and settlement will take place.

(3) The thorough compacting of the natural soil by rolling with a roller of proper weight and shape until there is formed a uniform and unyielding surface.

(4) The placing on the natural soil so compacted of a thickness of an impervious and incompressible material sufficient to cut off all communication between the soil and the bottom of the pavement.

The character of the natural soil over which the roadway is to be built has an important bearing upon the kind of foundation and the manner of forming it; each class of soil will require its own special treatment. Whatever its character, it must be brought to a dry and tolerably hard condition by draining and rolling. Sand and gravels which do not hold water, present no difficulty in securing a solid and secure foundation; clays and soils retentive of water are the most difficult. Clay should be excavated to a depth of at least 8 inches below the bottom of the finished covering; and the space so excavated should be filled in with sand, furnace slag, ashes, coal dust, oyster shells, broken brick, or other materials which are not absorbent of water excessively. A clay soil or one retaining water may be cheaply and effectually improved by laying cross drains with open joints at intervals of 50 or 100 feet. These drains should be not less than 18 inches below the surface, and the trenches should be filled with gravel. They should be 4 inches in internal diameter, and should empty into longitudinal drains.

Sand and planks, gravel and broken stone successively have been used to form the foundation for pavements; but, although eminently useful materials, their application to this purpose always has been a failure. Being inherently weak and possessing no cohesion, the main reliance for both strength and wear must be placed upon the surface covering. This covering—usually (except in case of sheet asphalt) composed of small units, with joints between them varying from $\frac{1}{2}$ inch to $1\frac{1}{2}$ inches—posesses no elements of cohesion; and under the blows and vibrations of traffic the independent units or blocks will settle and be jarred loose. On account of their porous nature, the subsoil quickly becomes saturated with urine and surface waters, which percolate through the joints; winter frosts upheave them; and the surface of the street becomes blistered and broken up in dozens of places.

Concrete. As a foundation for all classes of pavement (broken stone excepted), hydraulic-cement concrete is superior to any other. When properly constituted and laid, it becomes a solid, coherent mass, capable of bearing great weight without crushing. If it fail at all, it must fail altogether. The concrete foundation is the most costly, but this is balanced by its permanence and by the saving in the cost of repairs to the pavement which it supports. It admits of access to subterranean pipes with less injury to the neighboring pavement than any other, for the concrete may be broken through at any point without unsettling the foundation for a considerable distance around it, as is the case with sand or other incoherent material; and when the concrete is replaced and set, the covering may be reset at its proper level, without the uncertain allowance for settlement which is necessary in other cases.

Thickness of Course. The thickness of the concrete bed must be proportioned by the engineer; it should be sufficient to provide against breaking under transverse strain caused by the settlement of the subsoil. On a well-drained soil, 6 inches will be found sufficient; but in moist and clayey soils, 12 inches will not be excessive. On such soils a layer of sand or gravel, spread and compacted before placing the concrete, will be found very beneficial.

The proportions of the ingredients required for the manufacture of concrete are ascertained by measuring the voids in each ingredient. The strongest concrete will be produced when the volume of cement is slightly in excess of that required to fill the voids in the sand, and the volume of the combined cement and sand exceeds by about 10 per cent the volume of the voids in the stone or other material used for the aggregate. Concrete frequently is mixed in the arbitrary proportions of 1 part of cement, 3 parts of sand, and 6 parts of stone, and although the results have been satisfactory, the proportions may not be the most economical.

The ingredients of the concrete should be thoroughly mixed with just sufficient water to produce a plastic mass, without any surplus water running from it. After mixing, the concrete should be deposited quickly in place, and brought to a uniform surface and thickness by raking; then tamped until the mortar flushes to the surface, then left undisturbed until set. The surface of concrete laid during dry, warm, weather should be protected from the drying action of the sun while the initial setting is in progress. This may be accomplished by sprinkling with water as frequently as the rate of evaporation demands or by covering it with a layer of damp sand, straw, hay, or canvas. During freezing weather it is customary to suspend the laying of concrete for the reason that alternate freezing and thawing disintegrate it.

Measuring Voids in the Stone and Sand. The simplest method for measuring the voids and one sufficiently accurate for the manufacture of concrete is the "pouring method" in which a suitable vessel of known capacity (usually one cubic foot) is filled with the material, in which it is desired to ascertain the voids. Water then is poured into the vessel until its surface is flush with the surface of the material. The water is measured, and its amount is considered to equal the total of the voids.

STONE=BLOCK PAVEMENTS

Stone blocks commonly are employed for pavements where traffic is heavy. The material of which the blocks are made should possess sufficient hardness to resist the abrasive action of traffic, and

TABLE XIII

Specific Gravity, Weight, Resistance to Crushing, and Absorptive Power of Stones

Material	Specific Gravity		Weight (lb. per cu. ft.)		Resistance to Crushing (lb. per sq. in.)		Percentage of Water Absorbed	
	Min.	Max.	Min.	Max.	Min.	Max.	Min.	Max.
Granite	2.60	2.80	163	176	12,000	35,000	0.066	0.155
Trap	2.86	3.03	178	189	19,000	24,000	0.000	0.019
Sandstone	2.23	2.75	137	170	5,000	18,000	0.410	5.480
Limestone	1.90	2.75	118	175	7,000	20,000	0.200	5.000
Brick, paving	1.95	2.55			10,000	20,000		

sufficient toughness to prevent them from being broken by the impact of loaded wheels. The hardest stones will not give necessarily the best results in the pavement, since a very hard stone usually wears smooth and becomes slippery. The edges of the block chip off, and the upper face becomes rounded, thus making the pavement very rough.

The stone sometimes is tested to determine its strength, resistance to abrasion, etc.; but, as the conditions of use are quite different from those under which it may be tested, such tests are seldom satisfactory. However, examination of a stone as to its structure, the closeness of its grain, its homogeneity, porosity, etc., may assist in forming an idea of its value for use in a pavement. A low degree of permeability usually indicates that the material will not be greatly affected by frost. For data see Table XIII.

Materials. *Granite.* Granite is employed more extensively for stone-block paving than is any other variety of stone; and because of this fact, the term "granite paving" is generally used as being synonymous with stone-block paving. The granite employed should be of a tough, homogeneous nature. The hard, quartz granites usually are brittle, and do not wear well under the blows of horses' feet or the impact of vehicles; granite containing a high percentage of feldspar will be injuriously affected by atmospheric changes; and granite in which mica predominates will wear rapidly on account of its laminated structure. Granite possesses the very important property of splitting in three planes at right angles to one another, so that paving blocks may readily be formed with nearly plane faces and square corners. This property is called the rift or cleavage.

Sandstones. Sandstones of a close-grained, compact nature often give very satisfactory results under heavy traffic. They are less hard than granite, and wear more rapidly, but do not become smooth and slippery. Sandstones are generally known in the market by the name of the quarry or place where produced as "Medina", "Berea", etc.

Trap Rock. Trap rock, while answering well the requirements as to durability and resistance to wear, is objectionable on account of its tendency to wear smooth and become slippery; it is also difficult to break into regular shapes.

Limestone. Limestone usually has not been successfully employed in the construction of block pavements, on account of its lack of durability against atmospheric influences. The action of frost commonly splits the blocks; and traffic shivers them, owing to the lamination being vertical.

Cobblestone Pavement. Cobblestones bedded in sand possess the merit of cheapness, and afford an excellent foothold for horses; but the roughness of such pavements requires the expenditure of a large amount of tractive energy to move a load over them. Aside from this, cobblestones are entirely wanting in the essential requisites of a good pavement. The stones being of irregular size, it is almost impossible to form a bond or to hold them in place. Under the action of the traffic and frost, the roadway soon becomes a mass of loose stones. Moreover, cobblestone pavements are difficult to keep clean, and very unpleasant to travel over.

Belgian=Block Pavement. Cobblestones were displaced by pavements formed of small cubical blocks of stone. This type of pavement was laid first in Brussels, thence imported to Paris, and from there taken to the United States, where it has been widely known as the "Belgian-block" pavement. It has been largely used in New York City, Brooklyn, and neighboring towns, the material being trap rock obtained from the Palisades on the Hudson River.

The stones, being of regular shape, remain in place better than cobblestones; but the cubical form (usually 5 inches in each dimension) is a mistake. The foothold is bad; the stones wear round; and the number of joints is so great that ruts and hollows are quickly formed. This pavement offers less resistance to traction than cobble-stones, but it is almost equally rough and noisy.

Granite=Block Pavement. The Belgian block gradually has been displaced by the introduction of rectangular blocks of granite. Blocks of comparatively large dimensions were employed at first. They were from 6 to 8 inches in width on the surface, from 10 to 20 inches in length, with a depth of 9 inches. They merely were placed in rows on the subsoil, perfunctorily rammed, the joints filled with sand, and the street thrown open to traffic. The unequal settlement of the blocks, the insufficiency of the foothold, and the difficulty of cleansing the street, led to the gradual development of the latest type of stone-block pavement, which consists of narrow, rectangular blocks of granite, properly proportioned, laid on an unyielding and impervious foundation, with the joints between the blocks filled with an impermeable cement.

Experience has proved beyond doubt that this latter type of pavement is the most enduring and economical for roadways sub-jected to heavy and constant traffic. Its advantages are many, while its defects are few.

Advantages.

(1)   Adapted to all grades.

(2)   Suits all classes of traffic.

(3)   Exceedingly durable.

(4)   Foothold, fair.

(5)   Requires but little repair.

(6)   Yields but little dust or mud.

(7)   Facility for cleansing, fair.

Defects.

(1)   Under certain conditions of the atmosphere, the surface of the pavement becomes greasy and slippery.

(2)   The incessant din and clatter occasioned by the movement of traffic is an intolerable nuisance; it is claimed by many physicians that the noise injuriously affects the nerves and health of persons who are obliged to live or do business in the vicinity of streets so paved.

(3)   Horses constantly employed upon it soon suffer from the continual jarring produced in their legs and hoofs, and quickly wear out.

(4) The discomfort of persons riding over the pavement is very great, because of the continual jolting to which they are subjected.

(5) If stones of an unsuitable quality are used—for example, those that polish—the surface quickly becomes slippery and exceedingly unsafe for travel.

Blocks. *Size and Shape.* The proper size of blocks for paving purposes has been a subject of much discussion, and a great variety of forms and dimensions are to be found in all cities.

For stability, a certain proportion must exist between the depth, the length, and the breadth. The depth must be such that when the wheel of a loaded vehicle passes over one edge of the upper surface of a block, the block will not tend to tip up. The resultant direction of the pressure of the load and adjoining blocks always should tend to depress the whole block vertically; where this does not happen, the maintenance of a uniform surface is impossible. To fulfill this requirement, it is not necessary to make the block more than 6 inches deep.

Width. The maximum width of blocks is controlled by the size of horses' hoofs. To afford good foothold to horses drawing heavy loads, it is necessary that the width of each block, measured along the street, shall be the least possible consistent with stability. If the width be great, a horse drawing a heavy load, attempting to find a joint, slips back, and requires an exceptionally wide joint to pull him up. It is therefore desirable that the width of a block shall not exceed 3 inches; or that four blocks, taken at random and placed side by side, shall not measure more than 14 inches.

Length. The length, measured across the street, must be sufficient to break joints properly, for two or more joints in line lead to the formation of grooves. For this purpose the length of the block should be not less than 9 inches nor more than 12 inches.

Form. The blocks should be well squared, and must not taper in any direction; sides and ends should be free from irregular projections. Blocks that taper from the surface downwards (wedge-shaped) should not be permitted in the work; but if any are allowed, they should be set with the widest side down.

Manner of Laying Blocks. The blocks should be laid in parallel courses, with their longest side at right angles to the axis of the street, and the longitudinal joints broken by a lap of at least 2 inches,

Figs. 79 and 80. The reason for this is to prevent the formation of longitudinal ruts, which would happen if the blocks were laid lengthwise. Laying blocks obliquely and "herringbone" fashion has

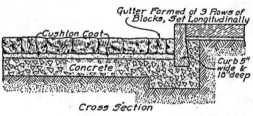

Fig. 79. Section Showing Method of Laying Stone-Block Pavement

been tried in several cities, with the idea that the wear and formation of ruts would be reduced by having the vehicle cross the blocks diagonally. The method has failed to give satisfactory results; the wear was irregular and the foothold defective; the difficulty of construction was increased by reason of labor required to form the triangular joints; and the method was wasteful of material.

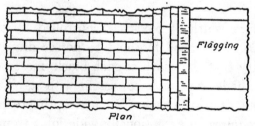

Fig. 80. Plan of Stone-Block Pavement Showing Method of Laying Blocks

The gutters should be formed by three or more courses of block, laid with their length parallel to the curb.

At junctions or intersections of streets, the blocks should be laid diagonally from the center, as shown in Fig. 81. The reasons for this are: (1) to prevent the traffic crossing the intersection from following the longitudinal joints and thus forming depressions and ruts; (2) laid in this manner, the blocks afford a more secure foothold for horses turning the corners. The ends of the diagonal blocks where they abut against the straight blocks, must be cut to the required bevel.

The blocks forming each course must be of the same depth, and no deviation greater than ¼ inch should be permitted. The blocks should be assorted as they are delivered, and only those corresponding in depth and width should be used in the same course. The better method would be to gage the blocks at the quarry. This would lessen the cost considerably; it would avoid also the inconvenience to the public due to the stopping of travel because of the rejection of defective material on the ground. This method undoubtedly would be preferable to the contractor, who would be saved the expense of handling unsatisfactory material; and it also would leave the inspectors free to pay more attention to the manner in which the work of paving is performed.

The accurate gaging of the blocks is a matter of much importance. If good work is to be executed, the blocks, when laid, must be in parallel and even courses; and if the blocks are not gaged accurately to one uniform size, the result will be a badly paved street, with the courses running unevenly. The cost of assorting blocks into lots of · uniform width, after delivery on the street, is far in excess of any

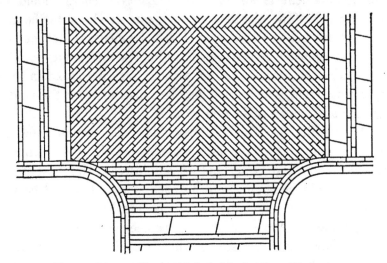

Fig. 81. Diagram Showing Method of Laying Stone Blocks at Intersection of Streets

additional price which would have to be paid for accurate gaging at the quarry.

Foundation. The foundation of the blocks must be solid and unyielding. A bed of hydraulic-cement concrete is the most suitable, and its thickness must be regulated according to the traffic; the thickness, however, should not be less than 4 inches, and need not be more than 9 inches. A thickness of 6 inches will sustain traffic of 600 tons per foot of width.

Cushion Coat. Between the surface of the concrete and the base of the blocks, there must be placed a cushion coat formed of an incompressible but mobile material, the particles of which readily will adjust themselves to the irregularities of the bases of the blocks and transfer the pressure of the traffic uniformly to the concrete below. A layer of dry, clean sand 1 inch to 2 inches thick forms an excellent cushion coat. Its particles must be of such fineness as to pass through a No. 8 screen; if the sand is coarse and contains pebbles, it will not adapt itself to the irregularities of the bases of the blocks; hence the blocks will be supported at a few points only, and unequal settlement will take place when the pavement is subjected to the action of traffic. The sand also must be perfectly free from moisture, and artificial heat must be used to dry it if necessary. This requirement is an absolute necessity. There should be no moisture below the blocks when laid; nor should water be allowed to penetrate below the blocks; if such happens, the effect of frost will be to upheave the pavement and crack the concrete.

Where the best is desired without regard to cost, a layer of asphaltic cement $\frac{1}{2}$ inch thick may be substituted for the sand, with superior and very satisfactory results.

Laying Blocks. The blocks should be laid stone to stone, so that the joint may be of the least possible width; wide joints cause increased wear and noise, and do not increase the foothold. The courses should be commenced on each side and should be worked toward the middle; and the last stone should fit tightly.

Ramming. After the blocks have been set, they should be well rammed down; and the stones which sink below the general level should be taken up and replaced with a deeper stone or brought to level by increasing the sand bedding.

The practice of workmen invariably is to use the rammer so as to secure a fair surface. This does not give the result intended to be secured, but brings each block to an unyielding bearing. The result of such a surfacing process is to produce an unsightly and uneven roadway when the pressure of traffic is brought upon it. The rammer used should weigh not less than 50 pounds and have a diameter of not less than 3 inches.

Fillings for Joints. All stone-block pavements depend for their waterproof qualities upon the character of the joint filling. Joints filled with sand and gravel of course are pervious. A grout of lime or cement mortar does not make a permanently waterproof joint; it becomes disintegrated under the vibration of traffic. An impervious joint can be made only by employing a filling made from bituminous or asphaltic material; this renders the pavement more impervious to moisture, makes it less noisy, and adds considerably to its strength.

Bituminous Cement for Joint Filling. The bituminous materials employed are: (1) coal tar having a specific gravity between 1.23 and 1.33 at 60 degrees Fahrenheit, a melting point between 120 and 130 degrees Fahrenheit, and containing not over 30 per cent of free carbon. (2) asphalt, either natural or artificial, entirely free from coal tar or any product of coal-tar distillation, and containing not less than 98 per cent of pure bitumen soluble in carbon bisulphide. Of the total amount soluble in carbon bisulphide, 98.5 per cent must be soluble in carbon tetrachloride. The penetration, when tested by the Dow method, must be not greater than 110, at 115 degrees Fahrenheit, and at 77 degrees Fahrenheit must range between 25 and 60. The specific gravity at 60 degrees Fahrenheit must not be more than 1.00.

The mode of applying the coal-tar filler is as follows: After the blocks are laid, gravel heated to about 250 degrees Fahrenheit is spread over the surface and swept into the joints until they are filled to a depth of about 2 inches. The blocks then are rammed. The coal-tar filler heated to a temperature between 250 and 300 degrees Fahrenheit is poured into the joints until they are about half filled, hot gravel is swept in until it reaches to within ½ inch of the surface, and hot filler is then poured in until it is flush with the surface of the blocks; after this sufficient hot gravel is applied to the joints to conceal the filler.

In applying the coal-tar filler it is essential that both the gravel and filler are heated sufficiently. Otherwise the filler will be chilled and will not flow to the bottom of the joint, but will form a thin layer near the surface, which under the action of frost and the vibration of traffic, will be cracked and broken up quickly; the gravel will settle, and the blocks will be jarred loose, causing the surface of the pavement to become a series of ridges and hollows. The filler should not be applied during a rainfall or while the blocks are wet or damp, for such a condition would prevent the filler from adhering to the blocks. The asphalt filler is heated to a temperature between 400 and 450 degrees Fahrenheit and poured into the joints until they are entirely filled.

Hydraulic-Cement Filler is composed of equal parts of. Portland cement and sharp sand mixed with clean fresh water to a suitable consistency. The joints between the blocks are filled to a depth of 2 inches with gravel, and the blocks are rammed, after which the filler is poured into the joints until they are filled flush with the surface of the blocks. In dry weather the blocks should be moistened by sprinkling with water before applying the filler. After the filler has taken its initial set, the whole surface of the pavement is covered with a layer of sand about ½ inch thick and if the weather is dry and warm it is sprinkled with water daily for three days. Traffic is not permitted to use the pavement until at least seven days after completion.

Stone Pavement on Steep Grades. Stone blocks may be employed on all practicable grades, but on grades exceeding 10 per cent, cobblestones afford a better foothold than blocks. The cobblestones should be of uniform length, the length being at least twice the breadth—say stones 6 inches long and 2½ inches to 3 inches in diameter.

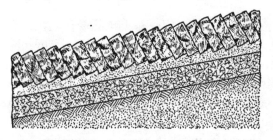

Fig. 82. Laying Stone Pavement on Steep Grades by Tilting Blocks

These should be set on a concrete foundation, laid stone to stone, and the interstices filled with cement grout or bituminous cement; or a bituminous-concrete foundation may be employed and the interstices between the stones may be filled with asphaltic paving cement. Should stone blocks be preferred, they must be laid, when the grade exceeds 5 per cent, with a serrated surface, by either of the methods shown in

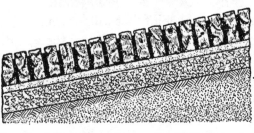

Fig. 83. Laying Stone Pavement on Steep Grades by Separating Blocks and Filling with Grout

Figs. 82 and 83. The method shown in Fig. 82 consists in slightly tilting the blocks on their bed so as to form a series of ledges or steps, which will insure a good foothold for horses' hoofs. The method shown in Fig. 83 consists in placing between the rows of stones a course of slate, or strips of creosoted wood, rather less than 1 inch in thickness and about 1 inch less in depth than the blocks; or the blocks may be spaced about 1 inch apart, and the joints filled with a grout composed of gravel and cement. The pebbles of the gravel should vary in size between $\frac{1}{4}$ inch and $\frac{3}{4}$ inch.

BRICK PAVEMENTS

A brick pavement consists of vitrified bricks laid on a suitable concrete foundation, Fig. 84.

Qualifications of Brick. The qualities essential to a good paving brick are the same as for any other paving material, viz, hardness, toughness, and ability to resist the disintegrating effects

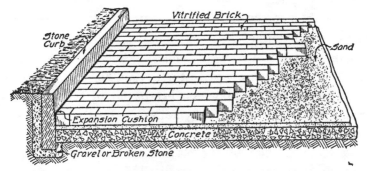

Fig. 84. Section Showing Method of Laying Vitrified Brick Pavement

of water and frost. These qualities are imparted to the brick by a process of annealing, through which the clay is brought to the point of fusion, and the heat then gradually reduced until the kiln is cold.

Composition. The material from which is made the majority of the brick used for paving is a shale. Shales are indurated clays with a laminated structure and the appearance of slate, and occur in stratified beds. The average composition of the shales that have proved satisfactory for the manufacture of paving brick is shown in Table XIV.

An excess of silica causes brittleness; or an excess of alumina causes shrinking, cracking, and warping. Iron renders the clay fusible and makes the brick more homogeneous. Lime in the form of silicate is valuable as a flux, but in the form of carbonate it will

TABLE XIV

Average Composition of Shales for Paving-Brick·Manufacture

Constituents	Proportional Part (per cent)
(Non-Fluxing)	
Silica	56.0
Alumina	22.0
Water and loss on ignition	7.0
Moisture	2.0
(Fluxing)	
Sesquioxide of iron	7.0
Lime	1.0
Magnesia	1.0
Alkalies	4.0
Total	100.0

decrease the strength of the brick; at a high temperature it is changed into caustic lime, which, while rendering the clay more fusible, will absorb moisture upon exposure to the weather and thus cause the brick to disintegrate. Magnesia exerts but little influence on the character of the brick. The alkalies in small quantities render the clay fusible.

Color. The color·of the clay is of no practical importance; it is due to the presence of the metallic oxides and organic substances. Iron produces bricks which are either red, yellow, or blue, according to the quantity present and the degree of heat; some organic substances produce a blue, bluish-gray, or black color.

The color of the brick is governed partly by the color of the clay, by the temperature of burning, by the kind of fuel used, and by the sand that is used to prevent the brick from sticking to the dies or to each other in the kiln.

Manufacture. In the manufacture of the brick, the shale is crushed usually in dry paws and then passed through a 4-mesh or an 8-mesh screen. The screened material is mixed with water in a pug mill to the required consistency. The finer the material is crushed and the more thoroughly it is worked or·tempered in the mill, the more uniform and better the brick is.

The plastic clay, in the "stiff-mud" process, as it leaves the pug mill is forced by an auger through a die which forms a bar of stiff clay of the desired dimensions, and this is cut by an automatic cutter into bricks of the size required. The bricks then, in some factories, are repressed in a die, during which the edges of the brick are rounded and the lugs, grooves, and trade-mark stamped on the sides. When repressing is not practiced, the bar of clay as it comes from the pug mill is cut by wires, the brick being called "wire-cut lug" brick.

The bricks, made by either method, are placed in a heated chamber to dry, this requiring from 18 to 60 hours according to the clay, temperature, and plant arrangement. When dry the bricks are stacked in the kiln, which is usually of the down-draft type with furnaces built in the outer walls. The bottom of the kiln is perforated to allow the gases to pass through to the flues placed below the floor and connected to the chimney. The heat from the furnace passes upward into the kiln, then downward through the bricks into the flues and thence to the chimney. At the beginning of the burning the heat is applied slowly to drive off the contained water without cracking the bricks. When the dryness of the smoke indicates the absence of moisture in the bricks, the fires are gradually increased until the temperature throughout the kiln is from 1500 to 2000 degrees Fahrenheit, this temperature being maintained from seven to ten days. The kiln then is closed, the fires are drawn, and the bricks are allowed to cool. This part of the process is called annealing, and to produce a tough brick requires from seven to ten days. The cooled bricks are sorted into different lots; the No. 1 paving bricks are generally found in the upper layers in the kiln.

Sizes. Two sizes of bricks are made: one size measuring $8\frac{1}{2} \times 2\frac{1}{2} \times 4$ inches weighing about 7 pounds and requiring 58 to the square yard. The other, measuring $8\frac{1}{2} \times 3\frac{1}{2} \times 4$ inches and frequently called "blocks", weighs about $9\frac{1}{2}$ pounds and requires 45 to the square yard.

Characteristics. The characteristics of brick suitable for paving are: not to be acted upon by acids—shale bricks not to absorb more than 2 per cent nor less than $\frac{1}{2}$ of 1 per cent of their weight of water, and clay bricks not to absorb more than 6 per cent of their weight of water (the absorption by a shale brick of less than $\frac{1}{2}$ of 1 per cent of its weight of water, indicates that it has been overburned); when broken with a hammer, to show a dense close-grained structure, free from lime, air holes, cracks, or marked laminations; not to scale, spall, or chip, when quickly struck on the edges; hard but not brittle.

Tests for Paving Brick. To ascertain if brick possesses the required qualities they are subjected to three tests: (1) abrasion by impact (commonly called the "rattler" test); (2) absorption; (3) cross breaking.

The Rattler Test. The rattler is a steel barrel 28 inches long and 28 inches in diameter, the sides formed of 14 staves fastened to two cast-iron heads furnished with trunnions which rest in a cast-iron frame. It is provided with gears and a belt pulley arranged to revolve at a rate of from $29\frac{1}{2}$ to $30\frac{1}{2}$ revolutions per minute. The material employed to abrade the brick is spherical balls of cast iron, the composition of which is: combined carbon, not less than 2.50 per cent; graphitic carbon, not more than 0.10 per cent; silicon, not more than 1 per cent; manganese, not more than 0.50 per cent; phosphorus, not more than 0.25 per cent; sulphur, not more than 0.08 per cent. Two sizes of balls or shot are used, the larger being 3.75 inches in diameter when new and weighing about $7\frac{1}{2}$ pounds, the smaller being 1.875 inches in diameter and weighing 0.95 pounds. A charge consists of ten large shot with enough small shot to make a weight of 300 pounds. The shot is used until the large size is worn to a weight of 7 pounds and the small shot is worn to a size that will pass through a circular hole $1\frac{3}{4}$ inches in diameter made in a cast-iron plate $\frac{1}{4}$-inch thick.

The brick to be tested are subjected to a temperature of 100 degrees Fahrenheit for three hours. Ten bricks are weighed and placed in the rattler with a charge of spherical shot, and the rattler is revolved for 1800 revolutions. The bricks then are taken out, pieces less than 1 pound in weight are removed and the balance weighed. From the weights before and after rattling the percentage of loss is calculated. The loss ranges from 16 per cent to 40 per cent. Brick to be used under heavy traffic should not lose more than 22 per cent, and for light traffic not more than 28 per cent.

Absorption Test. The absorption test is made on five bricks that have been through the rattler test. They are weighed, and are immersed in water for 48 hours, then are taken out and weighed, with the surplus water wiped off. From the weights before and after immersion the percentage of water absorbed is calculated.

Cross-Breaking Test. This test is made by placing a brick edge on supports 6 inches apart. The load is applied at the center of the brick, and is increased uniformly until fracture occurs. The average of the result on ten bricks is used in computing the modulus of rupture, $R = \dfrac{3WL}{2bd^2}$; in which W is the average breaking load in pounds, L the length between supports in inches, b the breadth, and d the depth in inches.

Brick=Pavement Qualifications. *Advantages.* The advantages of brick pavement may be stated as follows:

(1) Easy traction.
(2) Good foothold for horses.
(3) Not disagreeably noisy.
(4) Yields but little dust and mud.
(5) Adapted to all grades.
(6) Easily repaired.
(7) Easily cleaned.
(8) But slightly absorbent.
(9) Pleasing to the eye.
(10) Expeditiously laid.
(11) Durable under moderate traffic.

Defects. The principal defects of brick pavements arise from lack of uniformity in the quality of the bricks, and from the liability of incorporating in the pavement bricks too soft or too porous a structure, which crumbles under the action of traffic or frost.

Foundation. A brick pavement should have a firm foundation. As the surface is made up of small, independent blocks, each one must be supported adequately, or the load coming upon it may force it downwards and cause unevenness, a condition which conduces to the rapid destruction of the pavement. Several forms of foundation have been used—such as gravel, plank, sand, broken stone, and concrete. The last mentioned is the best.

Sand Cushion. The sand cushion is a layer of sand placed on top of the concrete to form a bed for the brick. Practice regarding the depth of this layer of sand varies considerably. In some cases it is only $\frac{1}{2}$ inch deep, varying from this up to 3 inches. The sand cushion is very desirable, as it not only forms a perfectly true and even surface upon which to place brick, but also makes the pavement less hard and rigid than would be the case were the brick laid directly on the concrete.

The sand is spread evenly, sprinkled with water, smoothed, and brought to the proper contour by screeds or wooden templets, properly trussed and mounted on wheels or shoes which bear upon the upper surface of the curb. Moving the templet forward levels and forms the sand to a uniform surface and proper shape.

The sand used for the cushion coat should be clean and free from loam, moderately coarse, and free from pebbles exceeding $\frac{1}{4}$ inch in size.

Manner of Laying. The bricks should be laid on edge or on one flat, as closely and compactly as possible, in straight courses across the street, with the length of the bricks at right angles to the axis of the street. Joints should be broken by at least 3 inches. None but whole bricks should be used, except in starting a course or making a closure. To provide for the expansion of the pavement, both longitudinal and transverse expansion joints are used, the former being made by placing a board templet $\frac{7}{8}$-inch thick against the curb and abutting the brick thereto. The transverse joints are formed at intervals varying between 25 and 50 feet, by placing a templet or building lath $\frac{3}{8}$-inch thick between two or three rows of brick. After the bricks are rammed and ready for grouting, these templets are removed, and the spaces so left are filled with coal-tar pitch or asphaltic paving cement. The amount of pitch or cement required will vary between 1 and $1\frac{1}{2}$ pounds per square yard of pavement, depending upon the width of the joints. After 25 or 30 feet of the pavement is laid, every part of it should be rammed with a rammer weighing not less than 50 pounds and the bricks which sink below the general level should be removed, sufficient sand being added to raise the brick to the required level. After all objectionable brick have been removed, the surface should be swept clean, then rolled with a steam roller weighing from 3 to 6 tons. The object of rolling is to bring the bricks to an unyielding bearing with a plane surface; if this is not done, the pavement will be rough and noisy and will lack durability. The rolling should be executed first longitudinally, beginning at the crown and working toward the gutter, taking care that each return trip of the roller covers exactly the same area as the preceding trip, so that the second passage may neutralize any careening of the brick due to the first passage.

The manner of laying brick at street intersections is shown in Fig. 85.

Joint Fillings. The character of the material used in filling the joints between the brick has considerable influence on the success

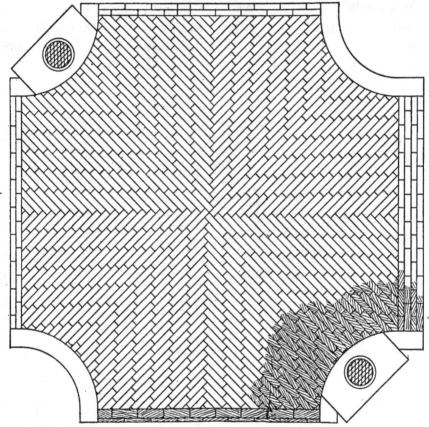

Fig. 85. Method of Laying Bricks at Street Intersections

and durability of the pavement. Various materials have been used—such as sand, coal-tar pitch, asphalt, mixtures of coal tar and asphalt, and Portland cement, besides various patented fillers, as "Murphy's grout", which is made from ground slag and cement. Each material has its advocates, and there is much difference of opinion as to which gives the best results.

The best results seem to be obtained by using a high grade of Portland cement containing the smallest amount of lime in its composition; the presence of the lime increasing the tendency of the filler to swell through absorption of moisture, causing the pavement to rise or to be lifted away from its foundation, and thus producing the roaring or rumbling noise so frequently complained of.

The Portland-cement grout, when uniformly mixed and carefully placed, resists the impact of traffic and wears well with brick. When a failure occurs, repairs can be made quickly, and, if made early, the pavement will be restored to a good condition. If, however, repairs are neglected, the brick soon loosen and the pavement fails.

Fig. 86. Grout Box Used in Laying Brick Pavement
Courtesy of National Paving Brick Manufacturers Association, Cleveland, Ohio

The office of a filler is to prevent water from reaching the foundation, and to protect the edges of the brick from spalling under traffic. In order to meet both of these requirements, every joint must be filled to the top, and must remain so, wearing down with the brick. Sand does not meet these requirements. Although at first making a good filler, being inexpensive and reducing the liability of the pavement to be noisy, it soon washes out, leaving the edges of the brick unprotected and consequently liable to be chipped. Coal tar and the mixtures of coal tar and asphalt have an advantage in rendering a pavement less noisy and in cementing together any breaks that may occur through upheavals from frost or other causes;

but, unless made very hard, they have the disadvantage of becoming soft in hot weather and flowing to the gutters and low places in the pavement, there forming a black and unsightly scale and leaving the high parts unprotected. The joints, thus deprived of their filling, become receptacles for water, mud, and ice in turn; and the edges of the brick are broken down quickly. Some of these mixtures become so brittle in winter that they crack and fly out of the joints under the action of traffic.

The Portland-cement filler is prepared by mixing 2 parts of cement and 1 part of fine sand with sufficient water to make a thin grout. The most convenient arrangement for preparing and distributing the grout is a water-tight wooden box carried on four wood wheels about 12 inches in diameter, Fig. 86. The box may be about 4 feet wide, 7 feet long, and 12 inches deep, furnished with a gate about 8 inches wide, in the rear end. The box should be mounted on the wheels with an inclination, so that the rear end is about 4 inches lower than the front end.

Following are the successive operations of placing the filler: The cement and sand are placed in the box, and sufficient water is added to make a thin grout. The grouting box is located about 12 feet from the gutter, the end gate opened, and about 2 cubic feet of the grout allowed to flow out and run over the top of the brick (care being taken to stir the grout while it is being discharged), Fig. 87. If the brick are very dry, the entire surface of the pavement should be wet thoroughly with a hose before applying the grout; if not, absorption of the water from the grout by the bricks will prevent adhesion between the bricks and the cement grout. The grout is swept into the joints by ordinary bass brooms. After a length of about 100 feet of the pavement has been covered the box is returned to the starting point, and the operation is repeated with a grout somewhat thicker than the first. If this second application is not sufficient to fill the joints, the operation is repeated as often as may be necessary to fill them. If the grout has been made too thin, or the grade of the street is so great that the grout will not remain long enough in place to set, dry cement may be sprinkled over the joints and swept in. After the joints are filled completely and inspected, allowing three or four hours to intervene, the completed pavement should be covered with sand to a depth

of about ½ inch, and the roadway barricaded, and no traffic allowed on it for at least ten days.

The object of covering the pavement with sand is to prevent the grout from drying or settling too rapidly; hence, in dry and windy weather, it should be sprinkled from time to time. If coarse sand is employed in the grout, it will separate from the cement during the operation of filling the joints, with the result that many joints will be filled with sand and very little cement, while others will be filled with cement and little or no sand; thus there will be

Fig. 88. Coal-Tar Heating Tank
*Courtesy of Barber Asphalt Paving Company,
Philadelphia, Pennsylvania*

many spots in the pavement in which no bond is formed between the bricks, and under the action of traffic these portions quickly will become defective.

The coal-tar filler is best applied by pouring the material from buckets, and brooming it into the joints with wire brooms; and in order to fill the joints effectually, it must be used only when very hot. To secure this condition, a heating tank on wheels is necessary, Fig. 88. It should have a capacity of at least 5 barrels, and be kept at a uniform temperature all day. One man is necessary to feed the fire and draw the material into the buckets; another, to carry

Fig. 89. Typical Brick Pavement Located in City. Eight Years Old When Picture Was Taken

the buckets from the heating tank to a third, who pours the material over the street. The latter starts to pour in the center of the street, working backward toward the curb, and pouring a strip about 2 feet in width. A fourth man, with a wire broom, follows immediately after him, sweeping the surplus material toward the pourer and in the direction of the curb. This method leaves the entire surface of the pavement covered with a thin coating of pitch, which immediately should be covered with a light coating of sand, the sand becoming imbedded in the pitch. Under the action of traffic, this thin coating is worn away quickly, leaving the surface of the bricks clean and smooth, Fig. 89.

Tools Used by Hand in the Construction of Block Pavements. The principal tools required in constructing block pavements comprise *hammers* and *rammers* of varying sizes and shapes, depending on the material and size of the blocks to be laid; also *crowbars*, sand *screens*, and rattan and wire *brooms*. Cobblestones, square blocks, and brick require different types both of hammer and rammers for adjusting them to place and for forcing them to their seats. A cobblestone rammer, for example, is usually made of wood (generally locust) in the shape of a long truncated cone, banded with iron at top and bottom, weighing about 40 pounds, and having two handles, one at the top and another on one side. A Belgian-block rammer is slightly heavier, consisting of an upper part of wood set in a steel base; while a rammer for granite blocks is still heavier, comprising an iron base with cast-steel face, into which is set a locust plug with hickory handles. For laying brick, a wooden rammer shod with cast iron or steel and weighing about 27 pounds is used. A light rammer of about 20 pounds weight, consisting of a metallic base attached to a long, slim, wooden handle, is used for miscellaneous work, such as tamping in trenches, next to curbs, etc.

Concrete=Mixing Machine. Where large quantities of concrete are required, as in the foundations of improved pavements, concrete can be prepared more expeditiously and economically by the use of mechanical mixers, and the ingredients will be mixed more thoroughly than by hand. Thorough incorporation of the ingredients is an essential element in the quality of a concrete. When mixed by hand, however, the incorporation is rarely complete, because it depends upon the proper manipulation of the hoe and shovel. The manipulation, although extremely simple, is rarely performed by the ordinary laborer as it should be unless he is watched constantly by the overseer.

Several varieties of concrete-mixing machines are in the market, all of which are efficient and of good design. A convenient portable

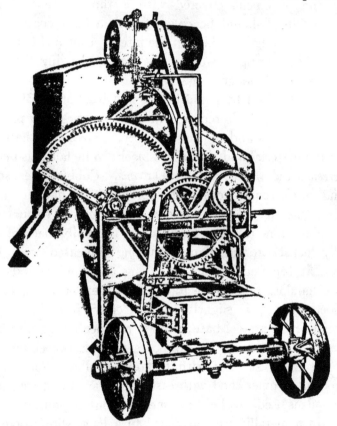

Fig. 90. Smith Concrete Mixer on Truck with Gasoline Engine,
Power Charger, and Water Tank
Courtesy of T. L. Smith Company, Milwaukee, Wisconsin

type is illustrated in Fig. 90. The capacity of the mixers ranges from 5 to 20 cubic yards per hour, depending upon size, regularity with which the materials are supplied, speed, etc.

Gravel Heaters. A special type of oven usually is employed for heating the gravel used for joint filling in stone-block pavements. These heaters are made in various sizes, a common size being 9 feet long, 5 feet wide, and 3 feet 9 inches high.

WOOD=BLOCK PAVEMENTS

Wood-block pavements, Fig. 91, are formed of rectangular blocks measuring from $3\frac{1}{2}$ to 4 inches wide, 5 to 10 inches long, and 4 inches deep, impregnated with creosote, or other preservative, laid in a bed of Portland-cement mortar spread upon a concrete foundation, with the joints between the blocks filled with either Portland-cement grout, or a bituminous filler.

The wood used is obtained from the long-leaved yellow pine (pinus palustrus), lob-lolly pine (pinus taeda), short-leaved pine (pinus echinata), Cuba pine (pinus heterophylla), black gum (nyssa sylvatica), red gum (liquidambar styrraciflua), Norway pine (pinus resinosa), or tamarack (larix laricina).

The wood should be cut from sound trees, free from cracks, shakes, and knots.

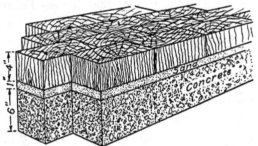

The great enemy of wood pavement is decay due to a low form of plant life called fungi. The fungi attack the wood from the outside, and if the wood is in the right condition for the spores to grow, they ulti-

Fig. 91. Section Showing Foundation for Wood-Block Pavements and Method of Laying Blocks

mately will penetrate the entire structure of the wood. There are three classes of fungi: one which attacks all parts of the wood structure; another which attacks the cellulose; and a third, which is the most common, and attacks only the lignin—the name of the many organic substances that are incrusted around the cellulose, and which with the latter constitute the essential part of woody tissue—here the fungi dissolve the lignin and the cellulose to make food for their development. Heat, air, and moisture are necessary to the existence of the fungi; without any one of these elements they cannot live. To destroy the fungus life and preserve wood from decay many processes have been devised; the one that seems to meet the requirements better than any other is the process of creosoting.

Creosoting. This process consists in impregnating the wood with the dead oil of tar, called "creosote", from which the ammonia has been removed. Its effect on the wood is to coagulate the albumen and thereby prevent its decomposition, also to fill the pores of the wood with a bituminous substance which excludes both air and moisture, and which is obnoxious to the lower forms of animal and vegetable life.

The coal-tar creosote oil is used without admixture or adulteration with other oils or tars. Its characteristics are: specific gravity, 1.03 to 1.08, at a temperature of 100 degrees Fahrenheit; contain not more than 5 per cent of tarry matter, nor more than 2 per cent of water, and not more than 8 per cent of tar acids, 99 per cent to be soluble in hot benzol; when subjected to distillation at gradually increasing temperatures up to 400 degrees Fahrenheit not more than 5 per cent of distillate shall pass over, at 450 degrees not more than 35 per cent, and up to 600 degrees Fahrenheit not more than 80 per cent; after complete distillation not more than 2 per cent of coke shall remain; upon sulphonating a sample of the total distillate, the residue shall not exceed 1 per cent.

For applying the creosote to the wood, several methods are followed. The one in most favor for paving blocks is the "pressure process", which essentially consists in: (1) steaming the wood for the purpose of liquefying the sap and other substances contained in the interfibrous spaces; (2) creating a vacuum for the purpose of removing the liquefied substances; (3) injecting the creosote under pressure.

The operation is performed in metal cylinders called "retorts", 6 or more feet in diameter and of any desired length, usually about 100 feet. The load of blocks, called a "charge", is placed upon metal cars called "buggies" and is run into the retort cylinder, the ends of which then are hermetically closed with "heads" or doors. Steam, at a gage pressure varying from 15 pounds to 45 pounds per square inch, is admitted to the retort (in some plants a vacuum is first created) and the pressure maintained for several hours. When the operator considers that the steaming has been continued a sufficient length of time, the products of condensation are removed from the retort through a blow-off cock in the bottom; when this is accomplished an air exhaust, or vacuum pump is put in operation, and a vacuum of from 20 inches to 26 inches is created and maintained for about one hour, at the end of which time the creosote is allowed to flow into the retort until it is filled. A pressure pump then is started to force the creosote into the retort until the pressure reaches 100 pounds to 150 pounds per square inch. This pressure is maintained until the required amount of creosote has been injected in the wood, then the surplus is drawn off, the heads opened, and the charge withdrawn.

The amount of creosote injected into the wood varies from 10 pounds to 22 pounds per cubic foot of wood. The amount is determined primarily by measuring the tanks and is verified by testing sample blocks. A sample block is bored entirely through in the direction of the fiber with an auger 1 inch in diameter, the hole being located midway between the sides and about $\frac{1}{3}$ the length of the block from one end. The borings are collected, thoroughly mixed, and the quantity and ratio of creosote to wood in the borings determined by extracting the creosote completely with carbon bisulphide.

The condition of the wood at the time of the treatment, is preferably dry and free from an excess of water. After treatment, and until used, the blocks during dry weather should be sprinkled frequently with water to prevent drying and cracking. The treated blocks are sometimes subjected to tests to determine the resistance to wear when saturated with water, the resistance to compression and impact, and to ascertain the amount of water the wood will absorb.

Laying the Blocks. The surface of the concrete foundation is cleansed from dust and dirt by sweeping, then sprinkled with water. Upon the cleaned surface a cushion coat is formed, by spreading a layer of sand 1 inch thick, Fig. 92, or a mortar composed of 1 part Portland cement and 2 parts sand, mixed with sufficient water to form a stiff paste (the practice of using a mixture of cement and sand slightly moistened with water produces a defective pavement). The blocks are set upon the cushion coat with the fiber vertical, Fig. 93, in straight, parallel courses at right angles to the axis of the street, except at intersections where they are set at an angle of 45 degrees with the axis of the street. They are laid so as to have the least possible width of joint (wide joints hasten the destruction of the wood by permitting the fibres to broom and wear under traffic). Blocks in adjoining courses break joint by at least 3 inches. At the

Fig. 92.   Spreading Sand Foundation for Wood Blocks in LaSalle Street, Chicago
Courtesy of Engineering News, New York City

Fig. 93.   Laying Wood Blocks in LaSalle Street, Chicago
Courtesy of Engineering News, New York City

Fig. 94. Wood-Block Pavement Being Hammered and Rolled, Preparatory to Putting in Filler
Courtesy of Engineering News, New York City

Fig. 95. Spreading Sand Filler on Wood-Block Pavement
Courtesy of Engineering News, New York City

curb it is customary to place one or two rows of blocks with the length parallel to the curb and ¾ inch therefrom.

After the blocks are laid they are brought to a uniform surface by ramming with hand rammers or rolling with a light steam roller, Fig. 94. When laid upon a mortar cushion, the rolling or ramming must be completed before the mortar sets.

In some cases the cushion coat is omitted, the surface of the concrete freed from dust by dry sweeping is covered with a thin coat of a bituminous cement and the blocks laid directly upon it. Sometimes, the side and one end of each block, when it is about to be set in place, are dipped in the same bituminous material that is used to cover the concrete, the blocks are placed in contact and the surface is covered with a thin coating of the bituminous material, this being covered with a layer of sand or fine gravel.

After the blocks have been brought to a uniform surface, the joints are filled with either fine sand, cement grout, or a bituminous cement, Fig. 95. When sand is used, it should be fine and dry, spread over the surface of the pavement, and swept about until the joints are filled. Cement grout is made of equal parts of Portland cement and fine sand mixed with water to the required consistency. It is spread over the surface of the blocks and swept into the joints until they are filled. The surface of the pavement then is covered with sand, and the grout is allowed to set for about seven days before traffic is admitted. The bituminous filler is composed of coal-tar pitch, asphalt, or combinations of these, and other ingredients. The filler is applied hot in the same manner as described under brick pavement. To provide for the expansion of the blocks the joint next the curb is filled with bituminous filler.

Qualifications of Wood Pavements. *Advantages.* The advantages of wood pavement may be stated as follows:

(1)   It affords good foothold for horses.

(2)   It offers less resistance to traction than stone, and slightly more than asphalt.

(3)   It suits all classes of traffic.

(4)   It may be used on grades up to 5 per cent.

(5)   It is moderately durable.

(6)   It yields no mud when laid upon an impervious foundation.

(7)   It yields but little dust.

(8) It is moderate in first cost.

(9) It is not disagreeably noisy.

Defects. The principal objections to wood pavement are:

(1) It is difficult to cleanse.

(2) Under certain conditions of the atmosphere it becomes greasy and very unsafe for horses. This may be remedied by covering the surface with a thin layer of fine sand or gravel; a similar treatment will absorb the oil which exudes during warm weather.

(3) It is not easy to open for the purpose of gaining access to underground pipes, it being necessary to remove rather a large surface for this purpose, which has to be left a little time after being repaired before traffic again is allowed upon it.

ASPHALT PAVEMENTS

Sheet=Asphalt Pavement. Sheet asphalt is the name used to describe a pavement having a wearing surface composed of sand graded in predetermined proportions, of a fine material or filler, and of asphalt cement, all incorporated by mixing in a mechanical mixer, and laid upon a concrete foundation, the surface of the latter being covered with a thin layer of bituminous concrete called a "binder".

Asphalt Cement. This is prepared from solid bitumen, refined and fluxed with (1) the residuum from paraffine petroleum; (2) the residuum from asphaltic petroleum; (3) a mixture of paraffine and asphaltic petroleum residuums; (4) natural malthas, or is prepared from (5) solid residual bitumen produced in the distillation of asphaltic petroleums, and fluxed with residuum oil produced from the same material.

Refined asphalt is that freed from the combined water and accompanying inorganic and organic matter. By comparatively simple operations the several varieties of asphalt may be separated from their impurities. Two methods are employed for refining; one using steam and the other direct fire. In both methods the asphalt is placed in tanks and slowly heated until thoroughly melted, and during the melting the mass is agitated by a current of either air or dry steam. The method of using steam is superior to the fire method. In the latter method there always is danger of overheating, in addition to the formation of coke and the cracking of the hydrocarbons.

The varieties of asphalt known as gilsonite and grahamite, which are practically pure bitumen, do not require refining, but they are used to a very small extent in paving.

The greater part of the solid bitumen used for paving in the United States is obtained from the West Indies and South America. The more extensively used being that found at Trinidad, W. I., and at Bermudez, Venezuela. The asphalts known by the trade names "california" and "texaco" are produced by refining asphaltic oils, and may or may not require to be fluxed.

Fluxes are fluid oils and tars which are mixed with asphalt to produce a desired consistency. The refined asphalt is melted and the flux previously heated added to it, in the proportion required to produce the desired consistency. The mixture of asphalt and flux is agitated either by mechanical means or by a blast of air until the materials are thoroughly incorporated and the desired consistency is obtained.

Sand. The sand should be siliceous and so free from organic matter, mica, soft grains, and other impurities, that these will not amount to more than 2 per cent of its volume.

Fine Material or Filler. This consists of any sound stone, usually limestone or sand, pulverized to such fineness that the whole will pass the No. 50 sieve, and not more than 10 per cent will be retained on the No. 100 sieve, and at least 70 per cent will pass the No. 200 sieve. Portland cement sometimes is used instead of the pulverized stone.

The paving composition is prepared by heating the ingredients separately to a temperature between 300 and 350 degrees Fahrenheit, then incorporating them by mixing in a mechanical mixer. The hot sand is measured into the mixer, followed by the hot filler; these two materials are thoroughly mixed together, and the hot cement is added in such a way as to distribute it evenly over the mixed sand and filler; the mixing then is continued until the materials form a uniform and homogeneous mass, with the grains of sand completely covered with cement. A typical mixture is: sand 100 pounds; filler 17.5 pounds; bitumen in asphalt cement 17.5 pounds.

The proportions of the ingredients in the paving mixture are not constant, but vary with the climate of the place where the pavement is to be used, the character of the sand, and the amount and character of the traffic that will use the pavement. The ranges are indicated in the following data:

Data for Asphaltic Paving

Asphaltic Paving Mixture.

CONSTITUENTS	PER CENT
Asphalt cement	12 to 15
Sand	70 to 83
Stone dust	5 to 15

Weight of Material. A cubic yard of the prepared material weighs about 4500 pounds. One ton of refined asphaltum makes about 2300 pounds of asphalt cement, equal to about 3.4 cubic yards of surface material.

Wearing Surface per Cubic Yard of Material.

THICKNESS (inches)	AREA (sq. yd.)
2½	12
2	18
1½	27

Laying the Pavement. The hot paving mixture is hauled to the street and dumped at a place outside of the space in which it

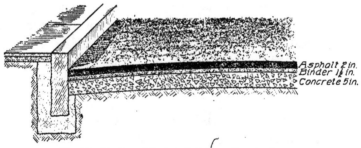

Asphalt 2 in.
Binder 1½ in.
Concrete 5 in.

Fig. 96. Section of Asphalt Pavement Showing Layers

is to be laid. It then is thrown into place with hot shovels, and spread with hot rakes uniformly to such a depth as will give the required thickness when compacted. The finished thickness varies between 1½ inches and 2 inches. The reduction of thickness by compression is about 40 per cent generally. Before the mixture is spread, the surfaces of curbs and street fittings that will be in contact with it are painted with hot asphalt cement.

The pavement is constructed in two forms: (1) The paving mixture is laid directly upon the surface of the concrete foundation;

(2) the surface of the concrete foundation is covered with a coat of asphaltic concrete, Fig. 96, called the "binder course", the object of which is to unite more securely the wearing surface to the foundation. This it does by containing a larger percentage of cement, which, if put in the surface mixture, would render it too soft. The binder is composed of sound, hard stone broken to pass a 1¼-inch screen, sand, pulverized stone, and asphalt cement, mixed in the desired proportions. A typical mixture is: stone 100 pounds; sand 40 pounds; stone dust 8 pounds; bitumen in asphalt cement 8 pounds.

The paving composition is compressed by means of rollers and tamping irons, the latter being heated in a fire contained in an iron basket mounted on wheels. These irons are used for tamping such portions as are inaccessible to the roller, namely, gutters, around manhole heads, etc.

Two rollers are sometimes employed; one, weighing 5 to 6 tons and of narrow tread, is used to give the first compression; and the other, weighing about 10 tons and of broad tread, is used for finishing. The rate of rolling varies; the average is about 1 hour for 1000 square yards of surface. After the primary compression, natural hydraulic cement, or any impalpable mineral matter, is sprinkled over the surface, to prevent the adhesion of the material to the roller and to give the surface a more pleasing appearance. When the asphalt is laid up to the curb, the surface of the portion forming the gutter is painted with a coat of hot cement.

Although asphaltum is a poor conductor of heat, and the cement retains its plasticity for several hours, occasions may and do arise through which the composition before it is spread has cooled; its condition when this happens is analogous to hydraulic cement which has taken a "set", and the same rules which apply to hydraulic cement in this condition should be respected in regard to asphaltic cement.

If the temperature of the air at the time of hauling is below 70 degrees Fahrenheit the wagons carrying it are covered with canvas or other material to prevent the loss of heat. The temperature when delivered at the place where it is to be used must not be less than 280 degrees Fahrenheit.

Two methods are followed in laying an asphalt pavement adjoining street railway tracks: (1) a course of granite-block or brick paving is laid between the rail and the edge of the asphalt; (2) the asphalt is laid directly against the rail, which, if its temperature is below 50 degrees Fahrenheit, is heated by suitable apparatus to a temperature of about 60 degrees Fahrenheit immediately before the asphalt is laid.

Foundation. A solid, unyielding foundation is indispensable with all asphaltic pavements, because asphalt of itself has no power of offering resistance to the action of traffic, consequently it nearly always is placed upon a bed of hydraulic-cement concrete. The concrete must be set thoroughly and its surface dry before the asphalt is laid upon it; if not, the water will be sucked up and converted into steam, with the result that coherence of the asphaltic mixture is prevented, and, although its surface may be smooth, the mass is really honey-combed, so that as soon as the pavement is subjected to the action of traffic, the voids or fissures formed by the steam appear on the surface, and the whole pavement is broken up quickly.

Qualifications of Asphalt Pavements. *Advantages.* These may be summed up as follows:

(1) It gives easy traction.

(2) It is comparatively noiseless under traffic.

(3) It is impervious.

(4) It is easily cleansed.

(5) It produces neither mud nor dust.

(6) It is pleasing to the eye.

(7) It suits all classes of traffic.

(8) There is neither vibration nor concussion in traveling over it.

(9) It is laid expeditiously, thereby causing little inconvenience to traffic.

(10) Openings to gain access to underground pipes are easily made.

(11) It is durable.

(12) It is repaired easily.

Defects. These are as follows:

(1) It is slippery under certain conditions of the atmosphere.

The American asphalts are much less so than the European, on account of their granular texture derived from the sand. The difference is very noticeable; the European are as smooth as glass,

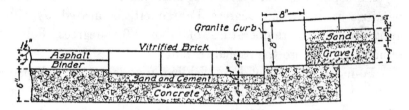

Fig. 97. Section of Asphalt Pavement Showing Use of Vitrified Brick to Form Gutter

while the American resemble fine sandpaper. The slipperiness can be decreased by heating the surface of the pavement with a surface heater, then applying a layer of coarse sand and rolling it into the surface.

(2) It will not stand constant moisture, and will disintegrate if sprinkled excessively.

(3) Under extreme heat it is liable to become so soft that it will roll or creep under traffic and present a wavy surface; and under extreme cold there is danger that the surface will crack and become friable.

(4) It is not adapted to grades steeper than $2\frac{1}{2}$ per cent, although it is in use on grades up to 7.30 per cent.

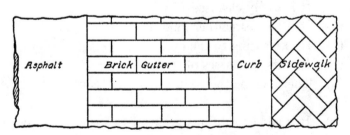

Fig. 98. Plan of Asphalt Pavement Showing Use of Vitrified Brick to Form Gutter

(5) Repairs must be made quickly, for the material has little coherence, and if, from irregular settlement of foundation or local violence, a break occurs, the passing wheels rapidly shear off the sides of the hole, and it soon assumes formidable dimensions.

Although pure asphaltum is impervious absolutely and insoluble in either fresh or salt water, yet asphalt pavements in the continued presence of water are quickly disintegrated. Ordinary rain or daily sprinkling does not injure them when they are allowed to become perfectly dry again. The damage is most apparent in gutters and adjacent to overflowing drinking fountains. This defect has long been recognized, and various measures have been taken to overcome it, or at least reduce it to a minimum. In some cities, ordinances have been passed, seeking to regulate the sprinkling of the streets; and in many places the gutters are laid with stone or vitrified brick, Figs. 97 and 98, while in others the asphalt is laid to the curb, a space of 12 to 15 inches along the curb being covered with a thin coating of asphalt cement.

Failure of Asphalt Pavement. The failure of asphalt pavement is due to any one, or a combination, of the following causes:

(1) *Unsuitable Materials.* The asphalt may be changed so by natural causes as to possess little or no cementing power. The fluxing agent may form only a mechanical instead of a chemical union with the asphalt, or its character may be such as to render the asphalt brittle, in which condition it easily is broken up under traffic. The sand may be graded improperly, either too coarse or too fine, or contain loam, vegetable matter, or clay.

(2) *Improper Manipulation.* The crude asphalt may have been refined at too high a temperature, which reduces or destroys the cementing property. The cement may be of improper consistence, of insufficient quantity, or inadequately mixed. If the cement is too hard, the pavement will have a tendency to crack during cold weather; and if too soft, it will push out of place and form waves under traffic. The quantity of cement needed varies with the character of the sand—a fine sand requires more cement than a coarse one, and the proportion of cement must be varied to suit the sand. When the ingredients are mixed inadequately, the cement and the particles of sand are not brought into intimate contact. Free oil or an excess of asphalt in the binder, making it too rich, is liable to work up and be absorbed by the wearing surface, and thus cause it to disintegrate. The mixture may be chilled while being transported from the plant to the street. There may be separation of the cement and sand, for if the distance from

the plant to the street is great and there is any delay, some of the cement may work to the bottom of the load, and when it is dumped, there will be fat and lean spots, both of which are injurious. The paving mixture may be laid upon a damp or dirty foundation. There may be inadequate compression, for the importance of thorough compaction is not appreciated always and this portion of the work is slighted frequently.

(3) *Natural Causes.* All materials in nature continually are undergoing change due to the action of the elements, and to this asphalt is not an exception. Subjected to the action of heat, all bitumens grow harder, and when the maximum degree of hardness is attained, natural decay sets in so that under the combined action of the elements, the material gradually rots and disintegrates.

(4) *Defective Foundation.* By unequal settlement a weak foundation will cause cracks and depressions in the surface which will enlarge speedily under traffic. A porous foundation will permit the ground water to rise, by capillary action, to the underside of the wearing surface, where by freezing it will cause cracks and thus provide access for surface water; non-watertight connection between curbs and street fittings also furnishes a path for surface water to reach the underside of the wearing surface, where the presence of water causes rapid decay.

(5) *Other Causes.* Illuminating gas, escaping from leaking pipes under the pavement causes disintegration of the asphalt. Contraction, caused by the decrease in cementing power through aging of the asphalt, results in cracks. Due to an excess of fluxing material, there may be rolling and waving of the pavement under traffic. Injury may be caused by fires made upon the pavement or by oil droppings from motor vehicles.

Sheet asphalt pavement usually is constructed under a contract that provides for its maintenance during a period of years (five or ten) by the contractor. Such a contract stipulates that the condition of the pavement at the expiration of the maintenance or guarantee period shall be as follows: Surface free from depressions exceeding $\frac{3}{4}$-inch deep, when measured between any two points 4 feet apart on a line conforming substantially to the original contour of the pavement. Free from cracks. Contain no disintegrated material. Thickness not reduced more than $\frac{3}{4}$ inch. Foundation free from cracks and settlement.

Rock Asphalt Pavement. This is the name applied to pavement made from the limestones and sandstones found naturally impregnated or cemented with bitumen. They are prepared for use by crushing and heating, and are used in their natural condition or mixed with other materials. Deposits are found in many parts of the United States and Europe. In Europe, rock asphalt is the material most extensively used for paving, under the name "asphalte". The European rock asphalts are impregnated very uniformly with from 7 per cent to 14 per cent of asphalt, and readily compact into a hard, smooth pavement which in frosty latitudes becomes very slippery. The American rock asphalts are impregnated irregularly with from 5 per cent to 30 per cent of asphalt. Their use for paving is limited, chiefly owing to the cost of transportation.

Asphalt Blocks. *Formation.* Asphalt paving blocks are formed from a mixture of asphaltic cement and crushed stone in the proportion of 8 or 12 per cent of cement to 88 or 92 per cent of stone. The materials are heated to a temperature of about 300° Fahrenheit, and mixed while hot in a suitable vessel. When the mixing is complete, the material is placed in molds and subjected to heavy pressure, after which the blocks are cooled suddenly by plunging into cold water. The usual dimensions of the blocks are 4 inches wide, 3 inches deep, and 12 inches long.

Foundation. The blocks usually are laid upon a concrete foundation with a cushion coat of sand about $\frac{1}{2}$-inch thick. They are laid with their lengths at right angles to the axis of the street, and the longitudinal joints should be broken by a lap of at least 4 inches. The blocks then are rammed with hand hammers, or are rolled with a light steam roller, the surface being covered with clean, fine sand; no joint filling is used, as, under the action of the sun and traffic, the blocks soon become cemented.

The advantages claimed for a pavement of asphalt blocks over those for a continuous sheet of asphalt are: (1) that they can be made at a factory located near the materials, whence they can be transported to the place where they are to be used and can be laid by ordinary paviors, whereas sheet pavements require special machinery and skilled labor; (2) that they are less slippery, owing to the joints and the rougher surface due to the use of crushed stone.

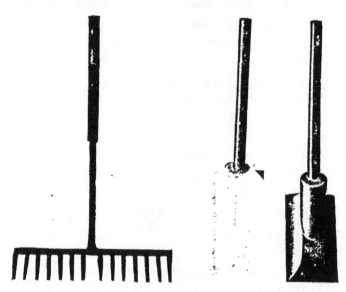

Fig. 99. Rake and Smoothing Irons Used in Laying Asphalt Pavement
Courtesy of Barber Asphalt Paving Company, Philadelphia, Pennsylvania

Another Form. Another form of asphalt block, known as the "Lueba" block, consists of a block 8¾ inches long, 4½ inches wide, and 4 inches thick, with the lower 3 inches composed of Portland-cement concrete covered with 1 inch of natural-rock asphalt; the two materials being joined under heavy hydraulic pressure. The blocks are laid on a concrete foundation and the joints between them are filled with hydraulic-cement grout.

Fig. 100. Pouring Pots Used with Asphalt Pavements
Courtesy of Barber Asphalt Paving Company

Tools Employed in Construction of Asphalt Pavements. The tools used in laying sheet-asphalt pavements comprise hand rammers iron, rakes, smoothing irons, Fig. 99; pouring pots, Fig. 100; hand rollers, either with or without a fire pot, Fig. 101; and steam rollers, with or without provision for heating the front roll, Fig. 102. These rollers are different in construction, appearance, and weight from those employed for compacting broken stone. The difference is due to the different character of the work required.

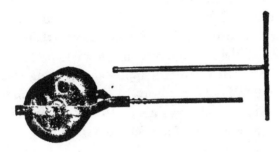

Fig. 101.   Hand Rollers Used in Laying Asphalt Pavements
Courtesy of Barber Asphalt Paving Company, Philadelphia, Pennsylvania

Fig. 102.   Small Road Roller Used in Laying Asphalt Pavements
Courtesy of Barber Asphalt Paving Company, Philadelphia, Pennsylvania

The principal dimensions of the 5-ton roller are as follows:

Front roll or steering wheel, diameter........	30 to 32 inches
Real roll or driving wheel, diameter.........	48 inches
Front roll, width..........................	40 inches
Rear roll, width.	40 inches
Length, extreme...........................	14 feet
Height, extreme.	7 to 8 feet
Water capacity............................	80 to 100 gallons
Coal capacity.............................	200 pounds

MISCELLANEOUS PAVEMENTS

Under this head will be described briefly the most notable examples of pavements devised as substitutes for the recognized standard types, and sometimes used where good materials are not available, or where insufficient funds prevent their purchase, and in some cases for the purpose of utilizing waste products.

Burnt Clay. In the Mississippi Valley, during the dry season, the clay is cut from the roadway to a depth of about 2 feet, and piled so as to form enclosures about 15 feet in diameter and 2 feet high. After remaining so for about ten days to dry out, a fire is made in the inclosure, more dry clay placed on top and the burning proceeded with. The burnt clay after cooling, is relaid upon the road, and then, being of a thoroughly porous nature, settles into a dry, solid layer.

Straw. Clay roads have been improved by shaping and harrowing the road, then applying a layer of wheat straw, which is moistened with water, and cut and mixed with the clay by a disk harrow. More straw is added and the operation repeated, then compacted with a steam roller. The treatment is applied twice a year.

Oyster=Shell. The shells are spread on the road previously shaped and rolled. They crush readily and, possessing a high cementing quality, bind together to form a compact, smooth road surface, but owing to their softness, they are quickly ground to powder which is carried away readily by wind and rain water.

Chert. The siliceous material found overlying the red sandstone, which forms the covering of the red hematite iron ore in some of the Southern States, is used for both street and road paving. It is laid directly upon the earth surface or upon a prepared foundation, sprinkled, and compacted in the same manner as water-bound macadam.

Slag. The slag produced in the manufacture of iron and steel is used in various ways for paving. (1) It may be crushed to the desired sizes and used in the same manner as broken stone, laid in one or two courses, sprinkled and rolled. In some cases, a binder composed of quicklime is used; in others, a waste sulphite liquor is mixed with the water used for moistening it before rolling; and in others, it is mixed with coal-tar or other bituminous cement and formed into a pavement in the same manner as bituminous macadam. The pavement called "tarmac", a large amount of which has been used in England, is composed of slag, coal tar, rosin, and Portland cement. (2) The slag may be formed into blocks by casting in molds, which are used in the same manner as stone blocks. In this form they are called "scoria" or "slag" blocks.

Clinker. Where crematories are employed for the destruction of garbage about 33 per cent of the material remains after burning, in the form of clinker. This is broken up and ground to a fine powder, mixed with either a hydraulic or a bituminous cement, and pressed into blocks and slabs.

Petrolithic. Petrolithic paving is made by applying a bituminous oil to earth, sand, gravel, clay, or loam roads. The soil is plowed to a depth of at least 6 inches, pulverized by harrowing, and sprinkled with water. The bituminous oil is applied in one or two coats at the rate of 1 gallon per square yard, the oil and soil are mixed and compacted by a roller weighing 5000 pounds, the surface of which is studded with spikes having a flat head measuring 2×3 inches, and on which account it is named a "sheep's-foot" roller. In operation, the spikes or feet are forced into the loose soil and compress or pack it from the bottom upwards. After a thorough mixing and tamping, the surface is shaped with a road grader and rolled with a roller of the ordinary form.

Kleinpflaster. Kleinpflaster is the name given to a stone pavement used in Germany for exceptionally heavy traffic, and used also in England, under the name "durax". It is made of 3-inch cubes of hard stone, cut by machinery, and laid in small segments of circles. The stones are laid as close as possible and the joints are filled with hydraulic-cement grout or bituminous filler.

Iron. Several experiments have been made with iron for paving, but, while eminently durable, it was rough, noisy, and slippery, and its use either alone or combined with other materials has been abandoned.

Trackways. Formed of stone slabs, brick, concrete blocks, steel, and other materials, trackways have been constructed at various times for the purpose of reducing the resistance to traction. Their use on an extensive scale, however, has been abandoned except in Italy, Spain, and Germany.

National Pavement. National pavement is composed of pulverized clay, loam, or ordinary soil, heated and mixed with liquid bitumen. The mixture is spread to a depth of 2 or 3 inches upon the surface of the compacted and drained natural soil and is compressed by a power roller.

Fibered Asphalt Pavement. Fibered asphalt pavement is composed of wood fiber, obtained as a waste product from the process of extracting tannin and asphalt. The fiber is heated and mixed with a predetermined quantity of asphalt. The hot mixture is run into molds forming small blocks which are shipped to the place of use. The blocks are there heated to a temperature of 275° F. in a traveling heater that moves along the roadway and from which the hot mixture emerges in a continuous stream 18 inches wide and is deposited on the previously prepared foundation to a depth of 4 inches. After spreading, it is compressed to a thickness of 2 inches with a power roller.

Westrumite. Westrumite is an asphalt cement temporarily liquefied by emulsification. It is mixed cold with broken stone in an ordinary concrete mixer, spread on the foundation, and compacted by rolling. The evaporation of the vehicle leaves the asphalt cement as the binder.

· MISCELLANEOUS STREET WORK

FOOTPATHS

A footpath or walk is simply a road under another name—a road for pedestrians instead of one for horses and vehicles. The only difference that exists is in the degree of service required; but the conditions of construction that render a road well adapted to its object are very much the same as those required for a walk.

The effects of heavy loads such as traverse carriage-ways are not felt upon footpaths; but the destructive action of water and frost is the same in either case, and the treatment to counteract or resist these elements as far as practicable, and to produce permanency, must be the controlling idea in each case, and should be carried out upon a common principle. It is not less essential that a walk should be well adapted to its object than that a road should be; and it is annoying to find it impassable or insecure and in want of repair when it is needed for convenience or pleasure. In point of economy, there is the same advantage in constructing a footway skilfully and durably as there is in the case of a road.

Width. The width of footwalks (exclusive of the space occupied by projections and shade trees) should be ample to accommodate comfortably the number of people using them. In streets devoted entirely to commercial purposes, the clear width should be at least one-third the width of the carriage-way; in residential and suburban streets, a very pleasing result can be obtained by making the walk one-half the width of the roadway, and by devoting the greater part to grass and shade trees.

Cross Slope. The surface of footpaths must be sloped so that the surface water will readily flow to the gutters. This slope need not be very great; $\frac{1}{8}$ inch per foot will be sufficient. A greater slope with a thin coating of ice upon it, becomes dangerous to pedestrians.

Foundation. As in the case of roadways, so with footpaths, the foundation is of primary importance. Whatever material may be used for the surface, if the foundation is weak and yielding, the surface will settle irregularly and become extremely objectionable, if not dangerous, to pedestrians.

Surface. The requirements of a good covering for sidewalks are:

(1)   It must be smooth but not slippery.

(2)   It must absorb the minimum amount of water, so that it may dry rapidly after rain.

(3)   It must not be abrased easily.

(4)   It must be of uniform quality throughout, so that it may wear evenly.

(5)   It must neither scale nor flake.

(6)   Its texture must be such that dust will not adhere to it.

(7)   It must be durable.

Materials. The materials used for footpaths are as follows: stone, natural and artificial; wood; asphalt; brick; tar concrete; and gravel.

Stones. Of the natural stones, sandstone (bluestone) and granite are employed extensively. The bluestone, when well laid, forms an excellent paving material. It is of compact texture, absorbs water to a very limited extent, and hence soon dries after rain; it has sufficient hardness to resist abrasion, and wears well without becoming excessively slippery. Granite, although exceedingly durable, wears very slippery, and its surface has to be roughened frequently.

Slabs, of whatever stone, must be of equal thickness throughout their entire area; the edges must be dressed true to the square for the whole thickness (edges must not be left feathered as shown

Fig. 103.	Faulty Joint in Stone Sidewalk

in Fig. 103); and the slabs must be bedded solidly on the foundation and the joints filled with cement mortar. Badly set or faultily dressed flagstones are very unpleasant to walk over, especially in rainy weather; the unevenness causes pedestrians to stumble, and rocking stones squirt dirty water over their clothes.

Wood. Wood has been used largely in the form of planks; it is cheap in first cost, but proves very expensive from the fact that it lasts but a comparatively short time and requires constant repair to keep it from becoming dangerous.

Asphalt. Asphalt forms an excellent footway pavement; it is durable and does not wear slippery.

Brick. Brick of suitable quality, well and carefully laid on a concrete foundation, makes an excellent footway pavement for residential and suburban streets of large cities, and also for the main streets of smaller towns. The bricks should be good qualities of paving brick (ordinary building brick are unsuitable, as they soon wear out and are broken easily). The bricks should be laid in parallel rows on their edges, with their lengths at right angles to the axis of the path.

Concrete. Concrete or artificial stone is used extensively as a footway paving material. Its manufacture is the subject of several patents, and numerous kinds are to be had in the market. When manufactured of first-class materials and laid in a substantial manner, with proper provision against the action of frost, artificial stone forms a durable, agreeable, and inexpensive pavement.

Concrete walks are formed in one or two courses. In one-course work, the concrete is laid to a depth of 4 inches and tamped until sufficient mortar flushes to the surface to permit the forming of a smooth surface. In two-course work, the concrete for the base is spread and tamped to a depth of 3 inches, the top or surface course is spread upon the base before the latter has begun to set. The top course has a thickness of about 1 inch, and it is tamped and its surface is brought to the required plane by a straightedge and by troweling. Expansion is provided for by transverse joints extending the full depth of the concrete. The joints are placed 4 feet apart and are formed by placing across the side forms a $\frac{1}{4}$-inch thick metal dividing strip, which is removed before the cement sets so that the edges of the joint may be smoothed and rounded with a suitable tool.

The area to be covered by the walk is excavated to a minimum depth of 8 inches, or to such greater depth as the nature of the ground may require to secure a solid foundation. The surface of the ground so exposed is compacted by ramming, and a drainage course is formed of broken stone, gravel, or steam-plant cinders, thoroughly compacted by ramming, and its surface is brought to a plane parallel to and 4 inches below the finished surface of the concrete. In some situations it may be necessary to connect the drainage course with the sewers, street drains, or side ditches, for the purpose of furnishing an outlet for standing water; this is done by the use of 3-inch drain pipe placed where required.

The forms of steel or wood should be made substantially, and left in place until the concrete is set hard.

Concrete walks fail from the use of improper materials and defective workmanship, insufficient expansion joints, heaving and cracking by frost, due to imperfect drainage, displacement and cracks, due to settlement of the drainage course—this latter being frequent when cinders are used, as in time they are liable to decompose and shrink in volume and thus allow the concrete to settle. In two-course work failure may be in respect to flaking of the surface by the action of water and frost entering between and separating the courses. The concrete should not be laid during freezing weather, nor should frozen materials be used in the work.

CURBSTONES AND GUTTERS

Curbstones. Curbstones are employed for the outer side of footways, to sustain the covering and to form the gutter. Their upper edges are set flush with the footwalk pavement, so that the water can flow over them into the gutters

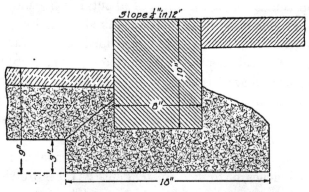

Fig. 104. Typical Section Showing Stone Curb Eight Inches Thick

The disturbing forces which the curb has to resist are: (1) The pressure of the earth behind it, which is frequently augmented by piles of merchandise, building materials, etc. This pressure tends to overturn it, break it transversely, or move it bodily on its base. (2) The pressure due to the expansion of freezing earth behind

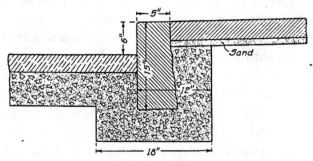

Fig. 105. Typical Section Showing Stone Curb Five Inches Thick

and beneath the curb. This force is most frequent where the sidewalk is sodded partly and the ground accordingly is moist. Successive freezing and thawing of the earth behind the curb will occasion a succession of thrusts forward, which, if the curb be of faulty design, will cause it to incline several degrees from the vertical.

(3) The concussions and abrasions caused by traffic. To withstand the destructive effect of wheels, curbs are faced with iron; and a concrete curb with a rounded edge of steel has been patented and used to some extent. Fires built in the gutters deface and seriously injure the curb. Posts and trees set too near the curb, tend to break, displace, and destroy it.

The use of drain tiles under the curb is a subject of much difference of opinion among engineers. Where the subsoil contains water naturally, or is likely to receive it from outside the curb lines, the use of drains is of decided benefit; but great care must be exercised in jointing the draintiles, lest the soil shall be loosened and removed, causing the curb to drop out of alignment.

The materials employed for curbing are the natural stones—as granite, sandstone (bluestone), etc.; artificial stone—fire clay, and cast iron.

The dimensions of curbstones vary considerably in different localities and according to the width of the footpaths; the wider the path, the wider should be the curb. However, it should be never less than 8 inches deep, nor narrower than 4 inches. Depth is necessary to prevent the curb's turning over toward the gutter. It never should be in smaller lengths than 3 feet. The top surface should be beveled off to conform to the slope of the footpath. The front face should be hammer-dressed for a depth of about 6 inches, in order that there may be a smooth surface visible against the gutter. The back for 3 inches from the top also should be dressed, so that the flagging or other paving may butt fair against it. The end joints should be cut a true square the full thickness of the stone at the top, and so much below the top as will be exposed; the remaining portion of the depth and bottom should be squared roughly, and the bottom should be fairly parallel to the top. (See Figs. 104 and 105.)

Combination Curb and Gutter. Concrete curb and gutter combined is constructed by placing the concrete in suitable forms. The concrete should be handled so as to prevent the separation of the stone and mortar, and when placed should be tamped well to bring the mortar to the surface and make complete contact with the forms. The corner formed by the top and face surfaces is rounded to a radius of about 1½ inches; sometimes this corner is formed of a steel bar put in place before the concrete is laid and anchored by metal strips spaced about 3 feet apart. Expansion joints are formed at distances of 10 or 12 feet. The remarks made under concrete walks regarding foundation, drainage, failure, etc., apply also to concrete curbs.

STREET CLEANING

The cleaning of streets is practiced for the purpose of protecting the health of the neighboring residents and for the comfort of the users. It is of comparatively recent development, and is rendered possible only by the use of hard pavements. The materials

Fig. 106. Typical Machine Street Sweeper
Courtesy of Acme Road Machinery Company, Frankfort, New York

to be removed from the streets consist of animal droppings, material worn from the pavement, materials dropped from vehicles, waste from building construction, miscellaneous materials swept from houses, stores, and factories, and the accumulation of snow during winter.

Cleaning Methods. The local conditions and character of the traffic and pavement determine the methods to be employed and the intervals for cleaning the streets. The methods employed are: sweeping, either by hand or by machine brooms; and flushing with water—the work being performed either during the day or the night, by large gangs at night, and by means of a patrol system during the day. Fig. 106 shows one of the machine sweepers used.

TABLE XV

Rate and Cost of Street Cleaning

Pavement	Approximate Surface Swept per Man (sq. yds. per hr.)		Approximate Dirt from Daily Sweeping (cu. yds. per 1000 sq. yds.)		Average Cost per Each Cleaning (cents per sq. yd.)
	(Wet)	(Dry)	(Min.)	(Max.)	
Asphalt	1000	1200	.007	.040	.0030
Granite-block	750	1000	.015	.024	.0050
Macadam (water-bound)	700		.100	.350	.0106
Wood			.070	.200	.0070
Brick					.0034

In the hand-cleaning method by day patrol, each man is furnished with a push broom, shovel, and can carrier in which to place the refuse, and has a certain section of street to clean each day. The day patrol sometimes is supplemented by a large gang working during the night. When machine brooms are employed they usually are operated at night and are supplemented by the patrol system during the day. As to which is the most economical, it will depend upon the cost of labor and the condition of the pavements; on pavements covered with ruts and depressions machine brooms are ineffective.

The approximate costs of the various methods of street cleaning per 1000 sq. yds. are:

Sweeping (hand)..................................$0.281
Sweeping (machine)............................. 0.317
Flushing (hand-hose)............................ 0.319
Flushing (machine).............................. 0.721

The average cost of supervision varies from .011 cent to 34 cents per mile.

The amount of surface cleaned by a machine broom depends upon the width of the broom, the power of the horses or other motive power, gradient, and condition of the surface. The wider the broom the less will be the cost. The average speed of travel is about 1½ miles per hour.

In Table XV are indicated the amount of surface which an average man will sweep per hour, depending upon the condition of the pavement—dry, wet, or muddy; relative amount of dirt

produced by the different pavements, if swept daily; and the average cost of cleaning different pavements.

Removal of Snow. The methods employed for keeping roads and streets passable during the period of snowfall varies according to the climatic conditions. In localities subject to heavy falls ' of snow, and continuous low temperature that retards the melting of the snow until spring, two methods are followed: (1) a narrow track is opened by a snowplow, through the center of the road, the snow being formed into long, narrow heaps on each side; (2) the snow is not removed, but is compacted by rolling with a light-weight wood or metal roller, Fig. 107. In localities having light falls and in the larger cities, the snow is pushed by plows or rotary brooms toward the gutters from where it is loaded into vehicles, hauled to a natural waterway and dumped, or in the absence of this it is placed in vacant lots and in some cases it is disposed of by dumping into the sewers through the manholes, but this must be done carefully, as there is liability of choking the sewer by the snow's consolidating. Light falls may be disposed of by the application of a stream of water to the surface of the street thereby washing the snow into the sewer. Many machines have been devised for melting the snow by the application of steam, hot air, etc., but none of them have been successful economically. In some cities the snow is melted by an application of rock salt which produces a thawing action when mixed with the snow by the traffic, the slushy mixture so formed is swept to the gutters by machine brooms and washed into the sewers by a stream of water from the hydrants. Objection is made to this method on account of the intense cold produced and its injurious effect upon the feet of pedestrians and on the hoofs of horses.

In order to cause the minimum of inconvenience to traffic it is necessary that the snow be removed from the streets as quickly as possible, therefore, it is customary, before the arrival of winter, to lay out the method and organization required and to make arrangements for the quick mobilization of the force needed for its removal. To accomplish this the city is divided into districts, in each of which there is established a headquarters and depot stocked with the necessary tools to execute the work in that district, and to which the laborers report when the snow commences to fall.

Street Sprinkling. Streets and roads are sprinkled with water for the purpose of abating dust and cooling the air. While water-bound macadam and earth surfaces must be sprinkled to abate the dust, a stone-block, brick, asphalt, or wood pavement will not require sprinkling if thoroughly cleansed and kept clean. On unclean and badly maintained pavements, sprinkling with water as usually performed converts the fine dust into a slime which renders all smooth pavements slippery, and in warm weather it becomes a prolific breeding place for disease germs, it clings to the feet and clothing of pedestrians, and, with its accompanying germs, is carried into buildings and dwellings.

The average cost of sprinkling per square yard is $0.009.

The systems followed for executing the work of street cleaning, snow removal, and sprinkling are: (1) by contract where the contractor furnishes all the tools and labor; (2) by contract for the labor only, the city furnishing the tools and machinery; (3) by the city, with its own staff and machinery.

SELECTING THE PAVEMENT

The problem of selecting the best pavement for any particular case is a local one, not only for each city, but also for each of the various parts into which the city is imperceptibly divided; and it involves so many elements that the nicest balancing of the relative values for each kind of pavement is required in arriving at a correct conclusion.

In some localities, the proximity of one or more paving materials determines the character of the pavement; while in other cases a careful investigation may be required in order to select the most suitable material. Local conditions always should be considered hence it is not possible to lay down any fixed rule as to what material makes the best pavement.

Qualifications. The qualities essential to a good pavement may be stated as follows:

(1) It should be impervious.

(2) It should afford good foothold for horses and adhesion for motor vehicles.

(3) It should be hard and durable, so as to resist wear and disintegration.

(4) It should be adapted to every grade.

(5) It should suit every class of traffic.

(6) It should offer the minimum resistance to traction.

(7) It should be noiseless.

(8) It should yield neither dust nor mud.

(9) It should be cleaned easily.

(10) It should be cheap.

Interests Affected. Of the above requirements, numbers (2), · (4), (5), and (6) affect the traffic and determine the cost of haulage by the limitations of loads, speed, and wear and tear of horses and vehicles. If the surface is tough or the foothold bad, the weight of the load a horse can draw is decreased, thus necessitating the making of more trips or the employment of more horses and vehicles to move a given weight. A defective surface necessitates a reduction in the speed of movement and a consequent loss of time; it increases the wear of horses, thus decreasing their life service and lessening the value of their current services; it also increases the cost of maintaining vehicles and harness.

Requirements, numbers (7), (8), and (9), affect the occupiers of adjacent premises, who suffer from the effect of dust and noise; they also affect the owners of said premises, whose income from rents is diminished where these disadvantages exist. Numbers (3) and (10) affect the taxpayers alone—first, as to the length of time during which the covering remains serviceable; and second, as to the amount of the annual repairs. Number (1) affects the adjacent occupiers principally on hygienic grounds. Numbers (7) and (8) affect both traffic and occupiers.

Problem Involved in Selection. The problem involved in the selection of the most suitable pavement consists of the following factors: (1) adaptability; (2) desirability; (3) serviceability; (4) comparative safety; (5) durability; (6) cost.

Adaptability. The best pavement for any given roadway will depend altogether on local circumstances. Pavements must be adapted to the class of traffic that will use them. The pavement suitable for a road through an agricultural district will not be suitable for the streets of a manufacturing center; nor will the covering suitable for heavy traffic be suitable for a pleasure drive or for a residential district.

General experience indicates the relative fitness of the several

TABLE XVI

Resistance to Traction on Different Pavements

KIND OF PAVEMENT	TRACTIVE RESISTANCE	
	Lb. per Ton	Fraction of the Load
Sheet-asphalt	30 to 70	$\frac{1}{67}$ to $\frac{1}{30}$
Brick	15 to 40	$\frac{1}{133}$ to $\frac{1}{50}$
Cobblestone	50 to 100	$\frac{1}{40}$ to $\frac{1}{20}$
Stone-block	30 to 80	$\frac{1}{67}$ to $\frac{1}{25}$
Rectangular wood-block	30 to 50	$\frac{1}{67}$ to $\frac{1}{40}$
Round wood-block	40 to 80	$\frac{1}{50}$ to $\frac{1}{25}$

materials as follows: for country roads, suburban streets, and pleasure drives—broken stone; for streets having heavy and constant traffic —rectangular blocks of stone, laid on a concrete foundation, with the joints filled with bituminous or Portland-cement grout; for streets devoted to retail trade, and where comparative noiselessness is essential—asphalt, wood, or brick. More recent experience indicates that concrete, when properly laid and reinforced at necessary points, may be employed to advantage for any pavement, both as base and as wearing surface.

Desirability. The desirability of a pavement is its possession of qualities which make it satisfactory to the people using and seeing it. Between two pavements alike in cost and durability, people will have preferences arising from the condition of their health, personal prejudices, and various other intangible influences, causing them to select one rather than the other in their respective streets. Such selections often are made against the demonstrated economies of the case, and usually in ignorance of them. Whenever one kind of pavement is more economical and satisfactory to use than is any other, there should not be any difference of opinion about securing it, either as a new pavement or in the replacement of an old one.

The economic desirability of pavements is governed by the ease of movement over them, and is measured by the number of horses or pounds of tractive force required to move over them a given weight—usually 1 ton. The resistance offered to traction by different pavements is shown in Table XVI.

Serviceability. The serviceability of a pavement is its quality of fitness for use. This quality is measured by the expense caused to the traffic using it—namely, the wear and tear of horses and vehicles, loss of time, etc. No statistics are available from which to deduce the actual cost of wear and tear.

The serviceability of any pavement in great measure depends upon the amount of foothold afforded by it to the horses—provided, however, that its surface be not so rough as to absorb too large a percentage of the tractive energy required to move a given load over it. Cobblestones afford excellent foothold, and for this reason are largely employed by horse-car companies for paving between the rails; but the resistance of their surface to motion requires the expenditure of about 40 pounds of tractive energy to move a load of 1 ton. Asphalt affords the least foothold; but the tractive force required to overcome the resistance it offers to motion is only about 30 pounds per ton.

Comparative Safety. The comparison of pavements in respect to safety, is the average distance traveled before a horse falls. The materials affording the best foothold for horses are as follows, stated in the order of their merit:

(1) Earth, dry and compact.
(2) Gravel.
(3) Broken stone (macadam).
(4) Wood.
(5) Sandstone and brick.
(6) Asphalt.
(7) Granite blocks.

Durability. The durability of pavement is that quality which determines the length of time during which it is serviceable, and does not relate to the length of time it has been down. The only measure of durability of a pavement is the amount of traffic tonnage it will bear before it becomes so worn that the cost of replacing it is less than the expense incurred by its use.

As a pavement is a construction, it necessarily follows that there is a vast difference between the durability of the pavement and the durability of the materials of which it is made. Iron is eminently durable; but, as a paving material, it is a failure.

The durability of a paving material will vary considerably with

TABLE XVII

✓ Terms of Life of Various Pavements

MATERIALS	TERMS OF LIFE (Years)
Granite-block	12 to 30
Sandstone	6 to 12
Asphalt	10 to 14
Wood	7 to 15
Limestone	1 to 3
Brick	5 to 15
Macadam	5 to ?

the condition of cleanliness observed. One inch of overlying dirt will protect the pavement most effectually from abrasion, and prolong its life indefinitely. But the dirt is expensive; it injures apparel and merchandise, and is the cause of sickness and discomfort. In the comparison of different pavements, no traffic should be credited to the dirty one. The life or durability of different pavements under like conditions of traffic and maintenance, may be taken as shown in Table XVII.

Cost. First cost or the cost of construction, is largely controlled by the locality of the place, its proximity to the particular material used, and the character of the foundation. The question of cost is the one which usually interests taxpayers, and is problably the greatest stumblingblock in the attainment of good roadways. The first cost usually is charged against the property abutting on the highway to be improved. The result is that the average property owner always is anxious for a pavement that costs little, because he must pay for it, not caring for the fact that cheap pavements soon wear out and become a source of endless annoyance and additional expense. Thus false ideas of economy usually have stood, and undoubtedly always will stand to some extent, in the way of realizing that the best is the cheapest.

The pavement which has cost the most is not always the best; nor is that which cost the least the cheapest; the one which is truly the cheapest is the one which makes the most profitable returns in proportion to the amount expended upon it. No doubt there is a limit of cost to go beyond which would produce no practical benefit; but it always will be found more economical to spend enough to secure the best results, and this always will cost less in the long run. One dollar well spent is many times more effective than one-half of the amount injudiciously expended in the hopeless effort to reach sufficiently good results. The cheaper work may look as well as the more expensive, for a time, but very soon may have to be done over again.

Economic Benefit. The economic benefit of a good roadway is comprised in the following: its cheaper maintenance, the greater facility it offers for traveling, thus reducing the cost of transportation; the lower cost of repairs to vehicles, and less wear of horses, thus increasing their term of serviceability and enhancing the value of their present service; the saving of time; and the ease and comfort afforded to those using the roadway.

Relative Economies. The relative economies of pavements—whether of the same kind in different condition, or of different kinds in like good condition—are determined sufficiently by summing their cost under the following headings of account:

(1) Annual interest upon first cost and sinking fund.

(2) Annual expense for maintenance.

(3) Annual cost for cleaning and sprinkling.

(4) Annual cost for service and use.

(5) Annual cost for consequential damages.

Interest on First Cost. The first cost of a pavement, like any other permanent investment, is measurable for purposes of comparison by the amount of annual interest on the sum expended. Thus, assuming the worth of money to be 4 per cent, a pavement costing $4 per square yard entails an annual interest loss or tax of $0.16 per square yard.

Cost of Maintenance. Under this head must be included all outlays for repairs and renewals which are made from the time when the pavement is new and at its best to a time subsequent, when, by any treatment, it is put again in equally good condition. The gross sum so derived, divided by the number of years which elapse between the two dates, gives an annual average cost for maintenance.

Maintenance means the keeping of the pavement in a condition practically as good as when first laid. The cost will vary considerably depending not only upon the material and the manner in which it is constructed, but upon the condition of cleanliness observed, and the quantity and quality of the traffic using the pavement.

The prevailing opinion that no pavement is a good one unless, when once laid, it will take care of itself, is erroneous; there is no such pavement. All pavements are being worn constantly by traffic and by the action of the atmosphere; and if any defects which appear are not repaired quickly, the pavements soon become unsatisfactory and are destroyed. To keep them in good repair, incessant attention is necessary, and is consistent with economy. Yet claims are made that particular pavements cost little or nothing for repairs, simply because repairs in these cases are not made, while any one can see the need of them.

Cost of Cleaning and Sprinkling. Any pavement, to be considered as properly cared for, must be kept dustless and clean. While circumstances legitimately determine in many cases that streets must be cleaned at daily, weekly, or semiweekly intervals, the only admissible condition for the purpose of analysis of street expenses must be that of like requirements in both or all cases subjected to comparison.

The cleaning of pavements, as regards both efficiency and cost, depends (1) upon the character of the surface; (2) upon the nature of the materials of which the pavements are composed. Block pavements present the greatest difficulty; the joints can never be perfectly cleaned. The order of merit as regards facility of cleansing, is: (1) asphalt, (2) concrete, (3) brick, (4) stone, (5) wood (6) macadam.

Cost of Service and Use. The annual cost for service is made up by combining several items of cost incidental to the use of the pavement for traffic—for instance, the limitation of the speed of movement, as in cases where a bad pavement causes slow driving and consequent loss of time; or cases where the condition of a pavement limits the weight of the load which a horse can haul, and so compels the making of more trips or the employment of more horses and vehicles; or cases where conditions are such as to cause greater wear and tear of vehicles, of equipment, and of horses. If a vehicle is run 1500 miles in a year, and its maintenance cost $30 a year, then the cost of its maintenance per mile traveled is 2 cents. If the value of a team's time is, say $1 for the legitimate time taken in going 1 mile with a load, and in consequence of bad roads it takes double that time, then the cost to traffic from having to use that mile of bad roadway is $1 for each load. The same reasoning applies to circumstances where the weight of the load has to be reduced so as to necessitate the making of more than one trip. Again, bad pavements lessen not only the life service of horses, but also the value of their current service.

Cost for Consequential Damages. The determination of consequential damages arising from the use of defective or unsuitable pavements, involves the consideration of a wide array of diverse circumstances. Rough-surfaced pavements, when in their best condition, afford a lodgment for organic matter composed largely of the urine and excrement of the animals employed upon the roadway. In warm and damp weather, these matters undergo putrefactive fermentation, and become the most efficient agency for generating and disseminating noxious vapors and disease germs, now recognized as the cause of a large part of the ills afflicting mankind. Pavements formed of porous materials are objectionable on the same, if not even stronger, grounds.

Pavements productive of dust and mud are objectionable, and especially so on streets devoted to retail trade. If this particular disadvantage be appraised at so small a sum per lineal foot of frontage as $1.50 per month, or 6 cents per day, it exceeds the cost of the best quality of pavement free from these disadvantages.

Rough-surfaced pavements are noisy under traffic and insufferable to nervous invalids, and much nervous sickness is attributable to them. To all persons interested in nervous invalids, this damage from noisy pavements is rated as being far greater than would be the cost of substituting the best quality of noiseless pavement; but there are, under many circumstances, specific financial losses, measurable in dollars and cents, dependent upon the use of rough, noisy pavements. They reduce the rental value of buildings and offices situated upon streets so paved—offices devoted to pursuits wherein exhausting brain work is required. In such locations, quietness is almost indispensable, and no question about the cost of a noiseless pavement weighs against its possession.

TABLE XVIII

Comparative Rank of Pavements

Characteristics		Variety								
Qualities	Value (per cent)	Asphalt (sheet)	Asphalt (block)	Concrete	Macadam (bituminous)	Macadam (water-bound)	Brick	Granite	Sandstone	Wood
Low tractive resistance	20	20.0	19.0	18.0	19.0	11.0	18.0	12.0	14.0	20.0
Service on grades	10	3.0	3.0	7.0	4.0	8.0	9.0	10.0	10.0	2.0
Non-slipperiness	5	1.5	2.5	4.0	2.5	4.5	3.5	3.5	5.0	2.0
Favorableness to travel	5	5.0	4.5	3.5	4.0	4.5	3.5	3.5	4.0	4.5
Sanitariness	10	10.0	9.0	7.0	8.0	3.0	8.0	6.0	7.0	9.0
Noiselessness	3	2.5	2.5	2.0	2.5	2.5	1.5	1.0	1.5	3.0
Minimum dust	3	2.5	2.5	2.0	2.0	1.0	2.0	1.5	2.0	2.0
Ease of cleaning	5	5.0	5.0	3.5	4.0	1.0	3.5	1.5	1.5	5.0
Acceptability	4	3.5	3.5	2.5	3.0	1.5	2.5	2.0	2.5	4.0
Durability	15	7.5	8.5	6.0	3.0	1.5	10.0	15.0	14.0	11.5
Ease of maintenance	5	3.5	4.0	3.0	3.0	2.5	4.0	4.5	5.0	5.0
Cheapness (first cost)	10	4.5	4.0	5.0	7.5	10.0	4.0	3.0	3.5	3.0
Low annual cost	5	1.5	2.5	3.0	3.5	1.0	4.5	5.0	5.0	5.0
Totals..............	100	70.0	70.5	66.5	66.0	52.0	74.0	68.5	75.0	76.0
Approximate first cost (dollars per sq. yd.)........		2.30	2.65	1.85	1.35	1.00	2.65	3.25	3.00	3.45

When an investigator has done the best he can to determine such a summary of costs of a pavement, he may divide the amount of annual tonnage of the street traffic by the amount of annual costs, and know what number of tons of traffic are borne for each cent of the average annual cost, which is the crucial test for any comparison, as follows:

(1) Annual interest upon first cost and sinking fund.......$
(2) Average annual expense for maintenance and renewal..
(3) Annual cost for custody (sprinkling and cleaning)......
(4) Annual cost for service and use.....................
(5) Annual cost for consequential damages...............
Amount of average annual cost ....
Annual tonnage of traffic................................
Tons of traffic for each cent of cost.....................

Gross Cost of Pavements. Since the cost of a pavement depends upon the material of which it is formed, the width of the roadway, the extent and nature of the traffic, and the condition of repair and cleanliness in which it is maintained, it follows that in no two streets is the endurance or the cost the same, and the difference between the highest and lowest periods of endurance and amount of cost is very considerable.

Comparative Rank of Pavements. In Table XVIII is given the rank of the various pavements in percentage, prorated from the values assigned in the first column to the desired qualities. The pavement ranking first in any given quality is given the full value for that quality, the others grading down from this value according to the extent to which they possess the desired quality. An examination of the table shows macadam to be the cheapest; least durable, and most difficult to maintain and cleanse; rather favorable to travel; comparatively low in sanitariness; and high in annual cost. While the table may be used as an aid in determining the most suitable pavement according to the factors that are susceptible of a numerical value, the values assigned must be modified by local conditions; first cost will necessarily vary in different localities, and certain factors will be more important in one locality than another.

Specifications. A specification or detailed description of the various works to be carried out always is attached to a contract, and is prepared before estimates are called for. The prominent points that are essential to the production of a specification that will fulfill its purpose properly, are: (1) description of the work; (2) extent of the work; (3) quality of the materials; (4) tests for the materials; (5) delivery of the materials; (6) character of the workmanship; (7) manner of executing the work.

Attention to these points and a clear and accurate description of each detail (leaving nothing to be imagined) not only will contribute materially to the rapid and efficient execution of the work, but will avoid any future misunderstanding. Every item of the work should be allotted a separate clause, for confusion must ensue when a single clause includes descriptions of several matters.

As a rule it is undesirable to insert in specifications any dimensions or weights that can be shown on the drawings. However, when it is necessary to insert them, words should be used instead of numerals; the use of numerals, and particularly decimal numbers, should be avoided, as there is a risk of having them set up incorrectly by the typesetter and overlooked in the proofreading. When a numeral is used it should be followed by the word or words indicating the numeral, placing the numeral in parenthesis.

Brevity, so far as it is consistent with completeness, should prevail, but the word "et cetera" should be excluded rigidly, and the matters covered by it should be defined clearly. Neither should important points of the work be dismissed with the direction that "the work shall be done to the satisfaction of the engineer". A direction of this kind usually implies that the engineer does not know what he wants, and therefore leaves the matter to the superior knowledge of the contractor—an attitude not very creditable to the former. The only really legitimate use of this phrase is in a general clause referring to the whole of the work.

The specifying of impracticable sizes of materials must be avoided as it causes unnecessary discussion and frequently leads to a charge for "extras".

A clause or phrase permitting the furnishing of alternative materials or workmanship should be excluded, because such permission affords ground for dispute and difference of opinion. On the other hand, specifying that certain articles manufactured by a particular firm shall be used should be avoided, as it suggests unfairness on the part of the engineer, and may create the idea that his selection is not without profit to himself.

With regard to the actual methods of carrying out the work, the contractor should not be tied to any particular means of effecting the required end, unless special circumstances require it, for, provided the materials and workmanship are satisfactory, it is better to allow the contractor to use his own discretion as to the manner of producing the required result.

While the standard and proper tests for the materials always should be stipulated, yet if they are carried to an extreme degree, as frequently happens, they defeat their own object. When it becomes impossible to carry out certain unreasonable demands, the alternative is to evade them as much as possible; and it must be borne in mind that the more stringent the demand, the greater the difficulty in enforcing it.

Contracts. A good, clear, and comprehensive contract is a difficult thing to write, but it should be "common sense" from beginning to end, and should be the joint production of both engineering and legal ability, neither sacrificing the one feature to the other.

The stipulations of the contract form the legal part of the document and are distinct from the technical description of the work to be done. The essential points are: (1) time of commencement; (2) time of completion; (3) manner and times of payment; (4) prices for which the work is to be performed; (5) measurements; (6) damages for noncompletion; (7) protection of persons and property during the prosecution of work; (8) such special stipulations as may be required for the particular work that is being contracted for.

It should be borne in mind that the contract and specifications when duly signed by the parties interested, is a legal document, which must be produced in court in the event of a dispute arising, therefore, it is of the utmost importance that it be written clearly in simple language, the clauses being arranged in logical sequence, and the descriptions made exact and complete without being needlessly verbose.

High-sounding phrases, and duplication of statements or information, should be avoided as tending to confusion. Specifications are seldom judged by literary standards of excellence, therefore, words may be repeated again and again if they express the meaning of the writer more clearly and forcibly than an alternative would do.

In the case of a lengthy contract and specification, a complete index with the clause and page numbers will be found an aid to finding quickly any required subject; cross references may sometimes be introduced with advantage.

INDEX.

INDEX

A

	PAGE
Asphalt pavements	153
asphalt blocks	161
failure	159
foundation	157
laying	155
qualifications	157
rock asphalt	161
sheet-asphalt	153
tools employed in constructing	162
Axle friction	11

B

Bituminous-macadam	99
amiesite	103
asphaltic petroleums	105
bitulithic	103
bituminous materials, definitions of	104
bituminous materials, tests for	106
cement, bituminous	104
construction	99
features	99
rock asphalt	103
Brick pavements	133
brick, qualifications of good	133
brick-pavement, qualifications of	137
fillings, joint	139
foundation	137
hand tools used	145
heaters, gravel	146
laying, manner of	138
machine, concrete-mixing	145
sand cushion	137
test	136
Broken-stone road	85
compacting stone	93
construction	85
macadam and telford roads, suppression of dust on	95
quality	88
rock, testing	89
shape and size of stones	92
species of stone	91
spreading stone	92
thickness of stones	92

C

PAGE

City streets and highways 113
    arrangement 113
    asphalt 153
    brick 168
    cleaning 172
    drainage 119
    foundations 121
    grades 114
    pavement, miscellaneous 164
    selecting 176
    stone-block 123
    street work, miscellaneous 166
    transverse contour or crown 118
    width 113
    wood 147

City streets, foundations for 121
    concrete 122

Concrete pavements 106
    bituminous surface, with 108
    block or cube 108
    construction 106
    joints, expansion 108
    materials 107
    reinforced-concrete 108

Country roads and boulevards 1
    location 12
    maintenance and improvement 110
    methods, preliminary construction 35
    nature-soil roads 74
    vehicles, resistance to movement of 1

Country roads and boulevards, drainage of 38
    ends from weather, protection of 42
    fall, for 40
    gutters 43
    hillside roads 45
    location 39
    materials 40
    outlets 43
    road gutters, inner and outer 45
    side ditches 43
    silt 42
    soils, nature of 39
    springs in cuttings, treatment of 44
    water breaks 46

Culverts 46
    arch 53
    box 52

Culverts (Continued)   PAGE
    design, factors in _____ 46
    earthenware pipe _____ 49
    functions _____ 46
    iron pipe _____ 51
    short span bridges used as _____ 53

E

Earthwork_____ 55
    classification _____ 57
    cuts and fills, balancing _____ 55
    embankments, methods of forming_____ 58
    prosecution of_____ 58
    shrinkage in _____ 57
    slopes, side _____ 55
    tools_____ 60

G

Grade problems _____ 33
    establishing a grade_____ 34
    level stretches_____ 34
    minimum_____ 33
    undulating_____ 33
Gravel roads_____ 83
    laying gravel_____ 84
    preparation of gravel_____ 83
    repair_____ 85

L

Location of roads_____ 12
    instruments_____ 14
    object of _____ 13
    points to consider_____ 13
    reconnoissance _____ 13
    selection, final_____ 22
        elements entering into_____ 22
        grade problems_____ 33
        gradient_____ 31
        location, final _____ 30
        mountain roads, treatment of _____ 26
        profile, construction of _____ 31
        repose, angle of_____ 32
        roads, alignment of _____ 27
        typical cases, treatment of_____ 23
    survey, preliminary_____ 15
        bridge site_____ 22
        features _____ 15
        map _____ 20
        memoir_____ 22
        topography _____ 15

M

	PAGE
Miscellaneous pavement	164
burnt clay	164
chert	164
clinker	165
iron	165
kleinpflaster	165
oyster-shell	164
petrolithic	165
slag	164
straw	164
trackways	165
Miscellaneous street work	166
curbstones and gutters	170
footpaths	166
foundation	167
materials	167
slope, cross	167
surface	167
width	167
Mountain roads, treatment of	26

N

Nature-soil roads	74
earth	74
sand	77
sand and gravel soils, application of oil to	78
sand-clay	77

R

Roads, maintenance and improvement of	110
broken-stone	109
improvement of existing roads	110
systems	110
traffic census	111
Roads with special coverings	79
foundations	79
materials	79
preparation	80
thickness	79
types	81
road covering, elements of	79
surfaces, wearing	82
bituminous-macadam	99
broken-stone	85
classification	82
concrete	100
function	82

Roads with special coverings PAGE
  surfaces, wearing (Continued)
    gravel - - - - 83
    thickness - - - - 82

S

Selecting pavement - - - - 176
  benefit, economic - - - - 181
  contracts - - - - 187
  cost of pavements, gross - - - - 184
  economics, relative - - - - 181
  interests affected - - - - 177
  problem involved in - - - - 177
  qualifications - - - - 176
  rank of pavements, comparative - - - - 185
  specifications - - - - 185
Stone-block pavements - - - - 123
  Belgian-block - - - - 125
  blocks, dimensions of - - - - 127
  blocks, manner of laying - - - - 127, 130
  cobble-stone - - - - 125
  cushion coat - - - - 129
  foundation - - - - 129
  granite-block - - - - 126
  joints, filling for - - - - 130
  materials - - - - 124
    granite - - - - 124
    limestone - - - - 125
    sandstones - - - - 125
    trap rock - - - - 125
  ramming - - - - 130
  stone pavement, steep grades on - - - - 132
Street cleaning - - - - 172
  methods - - - - 172
  snow, removal of - - - - 175
  sprinkling - - - - 176
Streets, drainage of - - - - 119
  catch basins - - - - 120
  gutters - - - - 119
  surface - - - - 119

T

Table
  different road materials, proportionate rise of center to width of carriageway - - - - 37
  effects of grades a horse can draw on different pavements - - - - 9
  grades, methods of designating - - - - 32
  gross loads for horse on different pavements - - - - 9
  life of various pavements, terms of - - - - 180

Table (Continued)

PAGE

loaded vehicles over inclined roads, data for_____ 11
paving-brick manufacture, average composition of shales for _____ 134
rank of pavements, comparative rank of_____ 184
resistance due to gravity on different inclinations_____ 5
resistance to traction on different pavements_____ 178
resistance to traction on road surfaces_____ 2
rise of pavement center above gutter for different paving materials ___ 119
stones, specific gravity, weight, resistance to crushing, and absorptive
    power of_____ 124
street cleaning, rate and cost of _____ 173
traction power of horses at different velocities_____ 7
tractive power with time, variation of _____ 8
traffic census_____ 112
wind pressure for various vehicles_____ 12

V

Vehicles, resistance to movement of_____ 1
air, resistance of_____ 12
friction, axle _____ 11
power and gradients, tractive _____ 7
traction, resistance to _____ 1

W

Wood-block pavements_____ 147
blocks, laying _____ 149
creosoting _____ 147
qualifications_____ 152

Printed in the USA
CPSIA information can be obtained
at www.ICGtesting.com
LVHW042058290823
756683LV00004B/16

9 781408 609095